Global Behavior of Nonlinear Difference Equations of Higher Order with Applications

T0214927

Mathematics and Its Applications

Managing Editor:

M. HAZEWINKEL

Centre for Mathematics and Computer Science, Amsterdam, The Netherlands

Volume 256

Global Behavior of Nonlinear Difference Equations of Higher Order with Applications

by

V.L. Kocic

and

G. Ladas
Department of Mathematics,
University of Rhode Island,
Kingston, Rhode Island, U.S.A.

KLUWER ACADEMIC PUBLISHERS
DORDRECHT / BOSTON / LONDON

Library of Congress Cataloging-in-Publication Data

Kocic, V. L., 1953-
    Global behavior of nonlinear difference equations of higher order
with applications / by V.L. Kocic and G. Ladas.
        p.   cm. -- (Mathematics and its applications ; v. 256)
    Includes bibliographical references and indexes.

    1. Nonlinear difference equations--Numerical solutions.
I. Ladas, G. E. II. Title. III. Series· Mathematics and its
applications (Kluwer Academic Publishers) ; v. 256.
QA431.K66 1993
515'.625--dc20
                                                        93-7605

ISBN 978-90-481-4273-6

Published by Kluwer Academic Publishers,
P.O. Box 17, 3300 AA Dordrecht, The Netherlands.

Kluwer Academic Publishers incorporates
the publishing programmes of
D. Reidel, Martinus Nijhoff, Dr W. Junk and MTP Press.

Sold and distributed in the U.S.A. and Canada
by Kluwer Academic Publishers,
101 Philip Drive, Norwell, MA 02061, U.S.A.

In all other countries, sold and distributed
by Kluwer Academic Publishers Group,
P.O. Box 322, 3300 AH Dordrecht, The Netherlands.

Printed on acid-free paper

Contents

Preface   ix

Acknowledgements   xi

1 Introduction and Preliminaries   **1**
1.1   An Overview 1
1.2   Oscillation and General Background 4
1.3   Stability and General Background 8
1.4   Periodicity and General Background 15
1.5   Global Attractivity for First-Order Equations 17
1.6   Some Basic Lemmas 19
1.7   Some Useful Theorems from Analysis 24
1.8   Notes 25

2 Global Stability Results   **27**
2.1   Global Asymptotic Stability of $x_{n+1} = x_n f(x_n, x_{n-1})$ 27
2.2   Permanence of $x_{n+1} = x_n f(x_n, x_{n-k_1}, \ldots, x_{n-k_r})$ 35
2.3   Global Attractivity of $x_{n+1} = x_n f(x_n, x_{n-k_1}, \ldots, x_{n-k_r})$ 39
2.4   Global Attractivity of $x_{n+1} = \alpha x_n + F(x_{n-k})$ 46
2.5   Global Stability of $x_{n+1} = \sum_{i=0}^{k} a_i x_{n-i} + (1 - A)F(\sum_{i=0}^{m} b_i x_{n-i})$ . . . 48
2.6   Notes 53

3 Rational Recursive Sequences   **55**
3.1   The Rational Recursive Sequence $x_{n+1} = (a + bx_n)/(1 + \sum_{i=0}^{k} a_i x_{n-i})$ 56
3.2   The Rational Recursive Sequence
     $x_{n+1} = (a + \sum_{i=0}^{k} a_i x_{n-i})/(b + \sum_{i=0}^{k} b_i x_{n-i})$ 59
3.3   Global Stability of $x_{n+1} = 1/(\sum_{i=0}^{k} b_i x_{n-i})$ 64
3.4   Global Stability of $x_{n+1} = (a + bx_n)/(A + x_{n-k})$ 67
3.5   Notes 74

4 Applications   **75**

  4.1 A Discrete Delay Logistic Model 75

  4.2 A Simple Genotype Selection Model 80

  4.3 Periodicity in a Simple Genotype Selection Model 87

  4.4 A Model of the Spread of an Epidemic 99

  4.5 Nicholson's Blowflies 105

  4.6 A Discrete Analogue of a Model of Haematopoiesis 111

  4.7 A Discrete Baleen Whale Model 119

  4.8 A Semidiscretization of a Delay Logistic Model 123

  4.9 A Discrete Analogue of the Emden-Fowler Equation 125

  4.10 Notes 132

5 Periodic Cycles   **133**

  5.1 An Invariance for $x_{n+1} = \frac{a + x_n + \cdots + x_{n-k+2}}{x_{n-k+1}}$ 135

  5.2 The Five–Cycle: $x_{n+2} = \frac{1 + x_{n+1}}{x_n}$ 138

  5.3 The Eight–Cycle: $x_{n+3} = \frac{1 + x_{n+2} + x_{n+1}}{x_n}$ 141

  5.4 Symmetric Periodic Sequences 144

  5.5 Notes 152

6 Open Problems and Conjectures   **153**

  6.1 The Rational Recursive Sequence $x_{n+1} = (a + bx_n)/(A + x_{n-1})$ 153

  6.2 The Rational Recursive Sequence $x_{n+1} = (a + bx_n^2)/(c + x_{n-1}^2)$ 155

  6.3 A Model for an Annual Plant 159

  6.4 The Dynamics of $x_{n+1} = Ax_n^{-2} + x_{n-1}^{-1/2}$ 162

  6.5 A Discrete Model with Quadratic Nonlinearity 166

  6.6 A Logistic Equation with Piecewise Constant Argument 168

  6.7 Discrete Epidemic Models 169

  6.8 Volterra Difference Equations 169

  6.9 Global Attractivity of $x_{n+1} - x_n + px_{n-k} = f(x_{n-m})$ 170

  6.10 Neural Networks 172

  6.11 The Fibonacci Sequence Modulo π 174

  6.12 Notes 176

Appendix   **177**

A The Riccati Difference Equation   **177**

B A Generalized Contraction Principle   **189**

C **Global Behavior of Systems of Nonlinear Difference Equations**   **195**
    C.1 A Discrete Epidemic Model 195
    C.2 A Plant-Herbivore System 197
    C.3 Discrete Competitive Systems 199

Bibliography   **205**

Subject Index   **223**

Author Index   **225**

PREFACE

Nonlinear difference equations of order greater than one are of paramount importance in applications where the $(n+1)^{st}$ generation (or state) of the system depends on the previous k generations (or states). Such equations also appear naturally as discrete analogues and as numerical solutions of differential and delay differential equations which model various diverse phenomena in biology, ecology, physiology, physics, engineering and economics.

Our aim in this monograph is to initiate a systematic study of the global behavior of solutions of nonlinear scalar difference equations of order greater than one. Our primary concern is to study the global asymptotic stability of the equilibrium solution. We are also interested in whether the solutions are bounded away from zero and infinity, in the description of the semicycles of the solutions, and in the existence of periodic solutions. This monograph contains some recent important developments in this area together with some applications to mathematical biology. Our intention is to expose the reader to the frontiers of the subject and to formulate some important open problems that require our immediate attention.

There are many books in the literature dealing with the theory of linear difference equations and with the local stability of nonlinear equations. There are also several books dealing with one dimensional dynamical systems. However, there are no books dealing systematically with the global behavior of solutions of nonlinear difference equations of order greater than one. It is our hope that this book will stimulate interest among mathematicians to develop further basic results on the global behavior of such equations. This book may also encourage people in applications to develop more realistic models which involve nonlinear difference equations of order greater than one.

With the use of a computer one can easily experiment with difference equations and one can easily discover that such equations possess fascinating properties with a great deal of structure and regularity. Of course all computer observations and predictions must also be proven analytically. Therefore this is a fertile area of research, still in its infancy, with deep and important results.

Chapter 1 contains some basic definitions and results which are used throughout the book. In this sense, this is a self-contained monograph and the main prerequisite that the reader needs to understand the material in this book and to be able to attack the open problems and conjectures is a good foundation in analysis.

In Chapter 2 we present some general results on the global asymptotic stability and global attractivity of quite general nonlinear difference equations of order greater than one.

Our goal in Chapter 3 is to gain some understanding of the dynamics of rational recursive sequences.

Chapter 4 deals with the global asymptotic stability and the oscillatory character of some discrete models, as well as some discrete analogues of continuous models taken from Mathematical Biology and Physics.

Chapter 5 deals with periodic cycles and with the general properties of rational recursive sequences with one term in the denominator.

In Chapter 6 we present some conjectures and open problems about some interesting types of difference equations.

The appendices contain material which although not in the mainstream of scalar nonlinear difference equations of higher order, are presented either for the sake of completeness or with the hope that they may inspire some useful generalizations.

At the end of every chapter we have included some notes and references about the material presented.

In addition to Chapter 6 which is entirely devoted to conjectures and open problems, in Chapters 2 - 6 we have included numerous research projects which, we hope, will stimulate a lot of interest and enthusiasm towards the development of a general theory with realistic applications.

Kingston, 1993   V. L. Kocic
                                                                   G. Ladas

ACKNOWLEDGEMENTS

This monograph is the outgrowth of lecture notes and seminars which were given at the University of Rhode Island during the last five years. We are thankful to Professors R. D. Driver, E. A. Grove, J. Hoag, A. Ivanov, J. Jaros, G. A. Kamenskii, R. Levins, J. Schinas, S. Schultz, Z. Wang, and S. Zhang, and to our graduate students M. Arciero, W. Briden, E. Camouzis, R. C. DeVault, J. H. Jaroma, C. Kent, S. Kuruklis, C. Qian, I. W. Rodrigues, and P. N. Vlahos for their enthusiastic participation and for offering useful suggestions which helped to improve the exposition. Special thanks are due to Professor E. A. Grove for proofreading in great detail the entire manuscript.

Finally we wish to express our warmest thanks to Dr. David Larner, head of Science and Technology Division of Kluwer Academic Publishers, for his enthusiastic support of our work.

Chapter 1

Introduction and Preliminaries

The first section in this chapter is a brief overview of what this book is about. In the remaining sections we gather together some general definitions and some preliminary material which will be useful throughout the book. In this way this is a self-contained monograph and the main prerequisite that the reader needs to understand the material in this book and to be able to attack the open problems and conjectures is a good foundation in analysis. The reader is advised to just glance at the definitions and results which are stated in Sections 1.2 - 1.7 and return for the details as they are needed in the sequel.

1.1 An Overview

Our aim in this book is to initiate a systematic study of the global behavior of solutions of nonlinear scalar difference equations of order greater than one. The equations which we study are, for the most part, of the form

$$x_{n+1} = F(x_n, x_{n-1}, \ldots, x_{n-k}), \ n = 0, 1, \ldots \qquad (1.1.1)$$

where k is a positive integer and F is a continuous function with nonnegative values. We will assume that Eq.(1.1.1) has a unique positive equilibrium $\bar{x}$ and our primary concern is to study the global asymptotic stability of $\bar{x}$. We are also interested in whether the solutions are bounded away from zero and infinity, in the description of the semicycles of the solutions, and in the existence of periodic solutions. This monograph contains some recent important developments in this area together with some applications to mathematical biology. Our intention is to expose the reader to the frontiers of the subject and to formulate some important open problems that require our immediate attention.

Some of the simple equations which have motivated our interest in this subject

1

are the following:

$$x_{n+1} = \frac{a + bx_n}{A + x_{n-k}}, \ n = 0, 1, \ldots \tag{1.1.2}$$

$$x_{n+1} = \frac{a + \sum_{i=0}^{k-1} b_i x_{n-i}}{x_{n-k}}, \ n = 0, 1, \ldots \tag{1.1.3}$$

$$x_{n+1} = x_n \exp(\alpha \frac{1 - x_{n-k}}{1 + x_{n-k}}), \ n = 0, 1, \ldots \tag{1.1.4}$$

$$x_{n+1} = x_n \exp(r(1 - \sum_{i=0}^{k} \alpha_i x_{n-i})), \ n = 0, 1, \ldots \tag{1.1.5}$$

When $k = 0$, Eq.(1.1.2) has applications in Optics and Mathematical Biology and is known in the literature as the Riccati difference equation. See Saaty [1, p.182]. The asymptotic behavior and the periodic character of solutions of Eq.(1.1.2) when $k = 0$ have been thoroughly investigated by Brand [1]. See also Appendix A.

When $a = 0$ and $b > A$, Eq.(1.1.2) has been proposed by Pielou in her books Pielou [1, p.22] and [2, p.79] as a discrete analogue of the delay logistic equation

$$\frac{dN(t)}{dt} = rN(t)[1 - \frac{N(t - \tau)}{P}], \ t \geq 0. \tag{1.1.6}$$

See also Section 4.1.

The special cases of Eq.(1.1.3),

$$x_{n+1} = \frac{1}{x_n}, \ n = 0, 1, \ldots \tag{1.1.7}$$

$$x_{n+1} = \frac{1 + x_n}{x_{n-1}}, \ n = 0, 1, \ldots \tag{1.1.8}$$

$$x_{n+1} = \frac{1 + x_n + x_{n-1}}{x_{n-2}}, \ n = 0, 1, \ldots \tag{1.1.9}$$

have the fascinating properties that all their positive solutions are periodic with periods 2, 5 and 8, respectively. Eq.(1.1.8) was discovered by Lyness [1] while he was working on a problem in Number Theory. This same equation also has applications in Geometry (see Leech [1]) and in frieze patterns (see Conway and Coxeter [1,2]). See also Sections 5.1 - 5.3.

Although for $k \geq 4$ and $a = b_1 = \cdots = b_{k-1} = 1$, the positive solutions of Eq.(1.1.3) are not all periodic with the same period, it is true, however, that every such solution is bounded from below and from above by positive constants. See also Section 5.1.

The difference equation (1.1.4) appears in genetics as a simple genotype selection model. See also Sections 4.2 and 4.3.

Finally when $k = 0$, Eq.(1.1.5) has various applications as a discrete delay logistic model. See May [1,3] and May and Oster [1]. Eq.(1.1.5) can also be derived from the delay logistic equation with piecewise constant arguments. See also Section 4.8.

There are many books in the literature dealing with the theory of linear difference equations and with the local stability of Eq.(1.1.1). See, for example, Agarwal [1], Brand [2], Kelley and Peterson [1], Lakshmikantham and Trigiante [1], LaSalle [1], and Mickens [1]. For the classical theory of difference equations see, for example, Boole [1], Geljfond [1], Jordan [1], Milne-Thompson [1], and Nörlund [1]. There are also several books dealing with one dimensional dynamical systems of the form

$$y_{n+1} = F(y_n), n = 0, 1, \ldots$$

where F is a function mapping an interval into itself. See, for example, Beardon [1], Collet and Eckman [1], Devaney [1], Gulick [1], Hale and Kocak [1], Mira [1], Preston [1], and Sandefur [1]. However, there are no books dealing systematically with the global behavior of solutions of Eq.(1.1.1) with $k \geq 1$.

Nonlinear difference equations of order greater than one, that is, higher-order dynamical systems, are of paramount importance in applications where the $(n+1)^{st}$ generation (or state) of the system depends on the previous k generations (or states). See, for example, Bergh and Getz [1], Brown [2], Clark [1], Fisher [1], Fisher and Goh [1], Goh and Agnew [1], Levin and May [1], May, Conway, Hassel and Southwood [1], and Smith [1]. Such equations also appear naturally as discrete analogues and as numerical solutions of differential and delay differential equations which model various diverse phenomena in biology, ecology, physiology, physics, engineering and economics. See, for example, Edelstein–Keshet [1], Freedman [1], Golub and Ortega [1], Hale and Kocak [1], Hoppensteadt [1], Kelley and Peterson [1], Lakshmikantham and Trigiante [1], and Murray [1].

The difference equations which result from various discretizations of differential equations are of the form

$$y_{n+1} - y_n + f(y_n, y_{n-1}, \ldots, y_{n-k}) = 0, \ n = 0, 1, \ldots. \qquad (1.1.10)$$

In analogy to delay differential equations, we may say that Eq.(1.1.10) is a **first order delay difference equation with delay** $k+1$. Of course, Eq.(1.1.10) can be put in the form of Eq.(1.1.1). However, it may be sometimes advantageous to think of Eq.(1.1.10) as being a "delay equation" and to apply to it techniques inspired by delay differential equations. See, for example Driver, Ladas and Vlahos [1], and Györi and Ladas [2, Chapter 7]. For various applications of delay difference equations see, for example, Brown [2], Clark [1], Cooke, Calef and Level [1], Fisher and Goh [1], Levin and May [1], May [4], Morimoto [1-3], and Pielou [1,2].

It is our hope that this book will stimulate interest among mathematicians to develop further basic results on the global behavior of nonlinear scalar difference equations of order greater than one. This book may also encourage people in applications to develop more realistic models which involve such nonlinear difference equations.

With the use of a computer one can easily experiment with difference equations and one can easily discover that such equations possess fascinating properties with a great deal of structure and regularity. Of course all computer observations and predictions must also be proven analytically. Therefore this is a fertile area of research, still in its infancy, with deep and important results which require our attention.

1.2   Oscillation and General Background

In this section we present the definitions of various concepts of oscillation and the statements of all the results in oscillation theory which we will need throughout this book.

Definition 1.2.1 *Oscillation*

(a) A sequence $\{x_n\}$ is said to **oscillate about zero** *or simply to* **oscillate** *if the terms x_n are neither eventually all positive nor eventually all negative. Otherwise the sequence is called* **nonoscillatory**. *A sequence $\{x_n\}$ is called* **strictly oscillatory** *if for every $n_0 \geq 0$, there exist $n_1, n_2 \geq n_0$ such that $x_{n_1} x_{n_2} < 0$.*

(b) A sequence $\{x_n\}$ is said to **oscillate about** $\bar{x}$ *if the sequence $\{x_n - \bar{x}\}$ oscillates. The sequence $\{x_n\}$ is called* **strictly oscillatory about** $\bar{x}$ *if the sequence $\{x_n - \bar{x}\}$ is strictly oscillatory.*

When we talk about the oscillation of a solution of Eq.(1.1.1) about $\bar{x}$, we will assume that $\bar{x}$ is an **equilibrium point** of Eq.(1.1.1); that is,

$$\bar{x} = F(\bar{x}, \ldots, \bar{x}).$$

A sequence which oscillates about zero consists of a "string" of non-negative terms followed by a string of negative terms, or vice versa, and so on. We will call these strings **positive** and **negative semicycles**, respectively.

When we study the oscillation of a solution about $\bar{x}$, the semicycles are defined relative to $\bar{x}$ and consist of strings of terms greater than or equal to $\bar{x}$ followed by strings of terms below $\bar{x}$, and so on. More precisely we make the following definitions about the semicycles of a solution $\{x_n\}$ of Eq.(1.1.1) relative to $\bar{x}$.

Definition 1.2.2 *Semicycles*
A **positive semicycle** *of a solution $\{x_n\}$ of Eq.(1.1.1) consists of a "string" of terms $\{x_l, x_{l+1}, \ldots, x_m\}$, all greater than or equal to $\bar{x}$, with $l \geq -k$ and $m \leq \infty$ and such that*

$$\text{either} \quad l = -k \quad \text{or} \quad l > -k \quad \text{and} \quad x_{l-1} < \bar{x}$$

and

$$\text{either} \quad m = \infty \quad \text{or} \quad m < \infty \quad \text{and} \quad x_{m+1} < \bar{x}.$$

A **negative semicycle** *of a solution $\{x_n\}$ of Eq.(1.1.1) consists of a "string" of terms $\{x_l, x_{l+1}, \ldots, x_m\}$, all less than $\bar{x}$, with $l \geq -k$ and $m \leq \infty$ and such that*

$$\text{either} \quad l = -k \quad \text{or} \quad l > -k \quad \text{and} \quad x_{l-1} \geq \bar{x}$$

and

$$\text{either} \quad m = \infty \quad \text{or} \quad m < \infty \quad \text{and} \quad x_{m+1} \geq \bar{x}.$$

The first semicycle of a solution of Eq.(1.1.1) starts with the term x_{-k} and is positive if $x_{-k} \geq \bar{x}$ and negative if $x_{-k} < \bar{x}$.

A solution may have a finite number of semicycles or infinitely many semicycles. A strictly oscillatory solution of Eq.(1.1.1) has infinitely many semicycles.

Let $\bar{x}$ be an equilibrium point of Eq.(1.1.1). Then the solution $\{x_n\}$ of Eq.(1.1.1) with $x_n = \bar{x}$ for all $n \geq -k$ is called a **trivial solution**. A trivial solution has only one semicycle, namely, the positive semicycle $\{\bar{x}, \bar{x}, \ldots\}$. This semicycle is called a **trivial semicycle**.

Here we mention some basic results for the oscillation of linear and nonlinear difference equations. The following results which are extracted from Györi and Ladas [2] will be useful in the sequel.

Theorem 1.2.1 *Consider the linear homogeneous difference equation*

$$x_{n+k} + \sum_{i=1}^{k} q_i x_{n+k-i} = 0, \quad n = 0, 1, \ldots \tag{1.2.1}$$

where k is a nonnegative integer and $q_1, \ldots, q_k \in \mathbf{R}$. Then the following statements are equivalent:

(a) Every solution of Eq.(1.2.1) oscillates.

(b) The characteristic equation of Eq.(1.2.1)

$$\lambda^k + \sum_{i=1}^{k} q_i \lambda^{k-i} = 0$$

has no positive roots.

Corollary 1.2.1 _Assume that $p \in \mathbb{R}$ and k is a nonnegative integer. Then every solution of the delay difference equation_

$$y_{n+1} - y_n + p y_{n-k} = 0, \ n = 0, 1, \ldots$$

oscillates if and only if

$$p \geq 1 \ \text{ if } \ k = 0$$

and

$$p > \frac{k^k}{(k+1)^{k+1}} \ \text{ if } \ k \geq 1.$$

Theorem 1.2.2 _For $i = 1, 2, \ldots, m$ assume that_

$$p_i \in (0, \infty) \ \text{ and } \ k_i \in \{0, 1, \ldots\} \ \text{ with } \ \sum_{i=1}^{m}(p_i + k_i) \neq 1.$$

Let $\{P_i(n)\}$ be sequences of positive numbers such that

$$\liminf_{n \to \infty} P_i(n) \geq p_i \ \text{ for } \ i = 1, 2, \ldots, m.$$

Suppose that the linear difference inequality

$$z_{n+1} - z_n + \sum_{i=1}^{m} P_i(n) z_{n-k_i} \leq 0, \ n = 0, 1, \ldots$$

has an eventually positive solution. Then the following statements are true:

(a) the equation

$$\lambda - 1 + \sum_{i=1}^{m} p_i \lambda^{-k_i} = 0$$

has a positive root.

(b) the difference equation

$$x_{n+1} - x_n + \sum_{i=1}^{m} p_i x_{n-k_i} = 0, \ n = 0, 1, \ldots \tag{1.2.2}$$

has a positive solution.

Theorem 1.2.3 *Assume that $p \in (0, \infty)$ and k is a nonnegative integer with $p+k \neq 1$. Let $f \in C[\mathbf{R}, \mathbf{R}]$ be such that*

$$uf(u) > 0 \quad \text{for} \quad u \neq 0, \quad \lim_{u \to 0} \frac{f(u)}{u} = 1,$$

and assume that there exists a $\delta > 0$ such that

$$\text{either} \quad f(u) \leq u \quad \text{for} \quad u \in [0, \delta]$$

$$\text{or} \quad f(u) \geq u \quad \text{for} \quad u \in [-\delta, 0].$$

Then every solution of the nonlinear difference equation

$$x_{n+1} - x_n + pf(x_{n-k}) = 0, \quad n = 0, 1, \ldots$$

oscillates if and only if every solution of the linearized equation

$$y_{n+1} - y_n + p y_{n-k} = 0, \quad n = 0, 1, \ldots$$

oscillates.

The following theorem from Kocic and Ladas [2] is a quite general linearized oscillation result for difference equations.

Theorem 1.2.4 *Consider the difference equation*

$$x_{n+1} - x_n + f(x_{n-k_1}, \ldots, x_{n-k_m}) = 0, \quad n = 0, 1, \ldots \qquad (1.2.3)$$

where $k_1, \ldots, k_m$ are nonnegative integers, and $f \in C[\mathbf{R}^m, \mathbf{R}]$ is such that

$$f(u_1, \ldots, u_m) \geq 0 \quad \text{for} \quad u_1, \ldots, u_m \in [0, \infty),$$

$$f(u_1, \ldots, u_m) \leq 0 \quad \text{for} \quad u_1, \ldots, u_m \in (-\infty, 0],$$

$$f(u, \ldots, u) = 0 \quad \text{if and only if} \quad u = 0.$$

Suppose also that there exists $\delta > 0$ such that f has continuous first partial derivatives $D_i f$ for all $u_1, \ldots, u_m \in [-\delta, \delta]$ and such that, for $i = 1, 2, \ldots, m$

$$D_i f(0, \ldots, 0) = p_i > 0, \quad \sum_{i=1}^{m}(p_i + k_i) \neq 1,$$

and

$$\text{either} \quad f(u_1, \ldots, u_m) \leq \sum_{i=1}^{m} p_i u_i \quad \text{for} \quad u_1, \ldots, u_m \in [0, \delta]$$

$$\text{or} \quad f(u_1, \ldots, u_m) \geq \sum_{i=1}^{m} p_i u_i \quad \text{for} \quad u_1, \ldots, u_m \in [-\delta, 0].$$

Then every solution of Eq.(1.2.3) oscillates if and only if every solution of the associated linear equation (1.2.2) oscillates.

1.3   Stability and General Background

Before we present some stability definitions for the difference equation

$$x_{n+1} = F(x_n, x_{n-1}, \ldots, x_{n-k}), \ n = 0, 1, \ldots \tag{1.3.1}$$

it is convenient, for brevity and generality, to rewrite Eq.(1.3.1) in vector form,

$$Y_{n+1} = G(Y_n), \ n = 0, 1, \ldots \tag{1.3.2}$$

where

$$Y_n = \begin{bmatrix} y_n^0 \\ y_n^1 \\ \vdots \\ y_n^k \end{bmatrix} = \begin{bmatrix} x_n \\ x_{n-1} \\ \vdots \\ x_{n-k} \end{bmatrix} \quad \text{and} \quad G(Y_n) = \begin{bmatrix} F(y_n^0, y_n^1, \ldots, y_n^k) \\ y_n^0 \\ \vdots \\ y_n^{k-1} \end{bmatrix}.$$

In the sequel we will assume, without further mention, that the initial point Y_0 and all iterates of Y_0 lie in the domain of G, and so Eq.(1.3.2) defines $\{Y_n\}$ for all $n \geq 0$. Sometimes for simplicity we will assume that G maps $\mathbf{R}^{k+1}$ into $\mathbf{R}^{k+1}$ when we just need to assume that G maps a subset of $\mathbf{R}^{k+1}$ into itself.

An **equilibrium** point of Eq.(1.3.1) is a point $\bar{x} \in \mathbf{R}$ such that $\bar{x} = F(\bar{x}, \bar{x}, \ldots, \bar{x})$. That is, $\bar{x}$ is a fixed point of the function $F(x, x, \ldots, x)$. Similarly, an equilibrium point of Eq.(1.3.2) is a vector $\bar{Y} \in \mathbf{R}^{k+1}$ such that $\bar{Y} = G(\bar{Y})$. Clearly, $\bar{x}$ is an equilibrium point of Eq.(1.3.1) if and only if

$$\bar{Y} = \begin{bmatrix} \bar{x} \\ \bar{x} \\ \vdots \\ \bar{x} \end{bmatrix}$$

is an equilibrium point of Eq.(1.3.2).

When Eq.(1.3.1) is linear of the form

$$x_{n+1} = b_0 x_n + b_1 x_{n-1} + \cdots + b_k x_{n-k}, \ n = 0, 1, \ldots \tag{1.3.3}$$

then by the above transformation, Eq.(1.3.3) becomes

$$Y_{n+1} = BY_n, \ n = 0, 1, \ldots \tag{1.3.4}$$

where B is the $(k+1) \times (k+1)$ matrix

$$B = \begin{bmatrix} b_0 & b_1 & \cdots & b_{k-1} & b_k \\ 1 & 0 & \cdots & 0 & 0 \\ \vdots & \vdots & \ddots & \vdots & \vdots \\ 0 & 0 & \cdots & 1 & 0 \end{bmatrix}. \tag{1.3.5}$$

In the sequel, we will denote by $\|\cdot\|$ any convenient vector norm and the associated matrix norm.

Definition 1.3.1 *Stability*

(a) *The equilibrium point $\bar{Y}$ of Eq.(1.3.2) is called* **stable** *(or* **locally stable***) if for every $\epsilon > 0$ there exists $\delta > 0$ such that $\|Y_0 - \bar{Y}\| < \delta$ implies $\|Y_n - \bar{Y}\| < \epsilon$ for $n \geq 0$. Otherwise the equilibrium $\bar{Y}$ is called* **unstable***.*

(b) *The equilibrium point $\bar{Y}$ of Eq.(1.3.2) is called* **asymptotically stable** *(or* **locally asymptotically stable***) if it is stable and there exists $\gamma > 0$ such that $\|Y_0 - \bar{Y}\| < \gamma$ implies*

$$\lim_{n\to\infty} \|Y_n - \bar{Y}\| = 0.$$

(c) *The equilibrium point $\bar{Y}$ of Eq.(1.3.2) is called* **globally asymptotically stable** *if it is asymptotically stable, and if for every Y_0,*

$$\lim_{n\to\infty} \|Y_n - \bar{Y}\| = 0.$$

(d) *The equilibrium point $\bar{Y}$ of Eq.(1.3.2) is called* **globally asymptotically stable relative to a set** $S \subset \mathbf{R}^{k+1}$ *if it is asymptotically stable, and if for every $Y_0 \in S$,*

$$\lim_{n\to\infty} \|Y_n - \bar{Y}\| = 0.$$

(e) *The equilibrium point $\bar{Y}$ of Eq.(1.3.2) is said to be a* **global attractor with basin of attraction** *a set $S \subset \mathbf{R}^{k+1}$ if*

$$\lim_{n\to\infty} Y_n = \bar{Y}$$

for every solution with $Y_0 \in S$.

Next, we will present the concept of Liapunov function. The following definition and theorem are from LaSalle [1].

Definition 1.3.2 *Liapunov function*
Let $S \subset \mathbf{R}^{k+1}$ be contained in the domain of G. A function $V : \mathbf{R}^{k+1} \to [0, \infty)$ is said to be a **Liapunov function** *of Eq.(1.3.2) on S if V is continuous and*

$$V(G(Y)) \leq V(Y) \quad \text{for all} \quad Y \in S.$$

Theorem 1.3.1 *Let $S \subset \mathbf{R}^{k+1}$ be a bounded open set such that $G(S) \subset S$. Assume that $\bar{Y}$ is the only equilibrium point of Eq.(1.3.2) in S. Suppose that V is a Liapunov function of Eq.(1.3.2) on S such that*

$$V(G(Y)) < V(Y) \quad \text{for all} \quad Y \in S \quad \text{with} \quad Y \neq \bar{Y}.$$

Let E_0 be the set of all points on the boundary of S such that $V(G(Y)) = G(Y)$, and let $M_0 \subset E_0$ be the set of all points such that $G(Y) = Y$ for $Y \in E_0$. Let $M = M_0 \cup \{\bar{Y}\}$. Then the following statements are true:

(a) If $M = \{\bar{Y}\}$, then $\bar{Y}$ is globally asymptotically stable relative to S.

(b) If there is no solution $\{Y_n\}$ of Eq.(1.3.2) such that

$$\inf_{Z \in M_0} \{\|Y_n - Z\|\} \to 0 \quad \text{when} \quad n \to \infty,$$

then $\bar{Y}$ is globally asymptotically stable relative to S.

The **characteristic equation** of Eq.(1.3.3) is

$$\lambda^{k+1} - b_0 \lambda^k - \cdots - b_k = 0 \tag{1.3.6}$$

which is also the characteristic equation of the matrix B in (1.3.5) and also the characteristic equation of the linear equation (1.3.4).

For a linear homogeneous equation, the stability of the zero equilibrium is equivalent to the boundedness of all solutions for $n \geq 0$. Also the asymptotic stability of the zero equilibrium is equivalent to **all** solutions having limit zero as $n \to \infty$ which in turn is true if and only if every root of the characteristic equation lies in the open unit disk $|\lambda| < 1$.

A linear equation will be called **stable, asymptotically stable,** or **unstable** provided that the zero equilibrium has that property.

The so-called **Schur-Cohn criterion** provides necessary and sufficient conditions for all roots of the equation

$$P(\lambda) = a_0 \lambda^n + a_1 \lambda^{n-1} + \cdots + a_{n-1} \lambda + a_n = 0 \tag{1.3.7}$$

with real coefficients to lie in the open unit disk $|\lambda| < 1$.

Before we can explain the Schur-Cohn criterion we need the so called **Routh–Hurwitz criterion.**

Theorem 1.3.2 *(Routh–Hurwitz criterion) Consider the polynomial equation (1.3.7) with real coefficients and $a_0 > 0$. Then a necessary and sufficient condition for all roots of Eq.(1.3.7) to have negative real parts is*

$$\Delta_k > 0 \quad \text{for} \quad k = 1, \dots, n \tag{1.3.8}$$

where Δ_k is the principal minor of order k of the $n \times n$ matrix

$$\begin{bmatrix} a_1 & a_3 & a_5 & \cdots & 0 \\ a_0 & a_2 & a_4 & \cdots & 0 \\ 0 & a_1 & a_3 & \cdots & 0 \\ \vdots & \vdots & \vdots & \ddots & \vdots \\ 0 & 0 & 0 & \cdots & a_n \end{bmatrix}.$$

Necessary and sufficient conditions for all roots of Eq.(1.3.7) to lie in the open unit disk $|\lambda| < 1$ are found from the Routh–Hurwitz criterion and the fact that the Möbius transformation

$$z = \frac{\lambda + 1}{\lambda - 1}$$

transforms the open unit disk in the λ–plane onto the open left-half plane in the z–plane.

Theorem 1.3.3 *(Schur-Cohn criterion) Eq.(1.3.7) has all its roots in the open unit disk $|\lambda| < 1$ if and only if the equation*

$$P\left(\frac{z+1}{z-1}\right) = 0$$

has all its roots in the left-half plane $\text{Re}(z) < 0$.

By applying the Schur-Cohn criterion we can easily obtain the following result for the asymptotic stability of the zero equilibrium of the second order difference equation

$$x_{n+2} + p x_{n+1} + q x_n = 0, \quad n = 0, 1, \ldots. \tag{1.3.9}$$

Theorem 1.3.4 *Assume $p, q \in \mathbf{R}$. Then a necessary and sufficient condition for the asymptotic stability of Eq.(1.3.9), is that*

$$|p| < 1 + q < 2. \tag{1.3.10}$$

The following result is the basic linearized stability theorem for nonlinear difference equations. For a proof see Lakshmikantham and Trigiante [1].

Theorem 1.3.5 *Consider the nonlinear difference equation*

$$X_{n+1} = A X_n + F(X_n), \quad n = 0, 1, \ldots \tag{1.3.11}$$

where A is a constant $k \times k$ matrix, $X_n \in \mathbf{R}^k$ for every $n \geq 0$, and $F \in \mathbf{C}[\mathbf{R}^k, \mathbf{R}^k]$ is such that

$$F(0) = 0 \quad \text{and} \quad \lim_{\|u\| \to 0} \frac{F(u)}{\|u\|} = 0. \tag{1.3.12}$$

Then the following statements hold:

(a) If all the eigenvalues of A lie in the open unit disk $|\lambda| < 1$, that is, if the linear equation

$$Y_{n+1} = AY_n, \; n = 0, 1, \ldots \tag{1.3.13}$$

is asymptotically stable, then Eq.(1.3.11) is also asymptotically stable.

(b) If at least one eigenvalue of A has absolute value greater than one, then Eq. (1.3.11) is unstable.

(c) If all eigenvalues of A lie in the closed unit disk $|\lambda| \leq 1$ and at least one eigenvalue has absolute value 1, then the stability of Eq.(1.3.11) cannot be determined by the mere stability of Eq.(1.3.13).

The local stability analysis for biological models which are described by some special scalar nonlinear delay difference equations were presented by Levin and May [1] and Clark [1] who established the following explicit conditions for local asymptotic stability.

Theorem 1.3.6 [Levin and May] *Assume that $q \in \mathbf{R}$ and $k \in \{0, 1, 2, \ldots\}$. Then the delay difference equation*

$$x_{n+1} - x_n + qx_{n-k} = 0, \; n = 0, 1, \ldots$$

is asymptotically stable if and only if

$$0 < q < 2\cos\frac{k\pi}{2k+1}.$$

A detailed proof of Theorem 1.3.6 can be found in Kuruklis [1] and Papanicolaou [1].

Theorem 1.3.7 [Clark] *Assume that $p, q \in \mathbf{R}$ and $k \in \{0, 1, \ldots\}$. Then*

$$|p| + |q| < 1$$

is a sufficient condition for the asymptotic stability of the difference equation

$$x_{n+1} + px_n + qx_{n-k} = 0, \; n = 0, 1, \ldots.$$

Remark 1.3.1 Theorem 1.3.7 can be easily extended to general linear equations of the form

$$x_{n+k} + p_1 x_{n+k-1} + \ldots + p_k x_n = 0, \; n = 0, 1, \ldots \tag{1.3.14}$$

where $p_1, \ldots, p_k \in \mathbf{R}$ and $k \in \{1, 2, \ldots\}$. Indeed, by applying Rouché's Theorem to the functions

$$f(z) = z^k \quad \text{and} \quad g(z) = p_1 z^{k-1} + \cdots + p_k$$

we can see that Eq.(1.3.14) is asymptotically stable provided that

$$\sum_{i=1}^{k} |p_i| < 1.$$

Remark 1.3.2 In Cooke and Györi [1] it was shown that the linear equation

$$x_{n+1} - x_n + \sum_{i=1}^{m} p_i x_{n-k_i} = 0, \ n = 0, 1, \ldots$$

where $p_1, \ldots, p_m \in (0, \infty)$ and $k_1, \ldots, k_m$ are positive integers, is asymptotically stable provided that

$$\sum_{i=1}^{m} k_i p_i < 1.$$

Theorem 1.3.5 deals with the stability of the zero equilibrium of Eq.(1.3.11). When (1.3.12) holds, Eq.(1.3.13) is called the **linearized equation** associated with Eq.(1.3.11). The linearized equation determines the local stability of the nonlinear equation in the sense of the statements (a) and (b) of Theorem 1.3.5.

Now suppose that we are interested in the local stability of the equilibrium point $\bar{X} \in \mathbf{R}^{k+1}$ of the vector difference equation

$$X_{n+1} = H(X_n), \ n = 0, 1, \ldots \tag{1.3.15}$$

where $X_n \in \mathbf{R}^{k+1}$ for every $n \geq 0$ and $H \in \mathbf{C}^1[\mathbf{R}^{k+1}, \mathbf{R}^{k+1}]$. Then by translating the equilibrium $\bar{X}$ to $0 \in \mathbf{R}^{k+1}$ one can see that the linearized equation associated with Eq.(1.3.15) is

$$Y_{n+1} = AY_n, \ n = 0, 1, \ldots$$

where A is the Jacobian matrix $DH(\bar{X})$ of the function H evaluated at the equilibrium $\bar{X}$.

Corollary 1.3.1 *Let $\bar{X}$ be an equilibrium point of Eq.(1.3.15) and assume that H is a $\mathbf{C}^1$ function in $\mathbf{R}^{k+1}$. Then the following statements are true:*

(a) If all the eigenvalues of the Jacobian matrix $DH(\bar{X})$ lie in the open unit disk $|\lambda| < 1$, then the equilibrium $\bar{X}$ of Eq.(1.3.15) is asymptotically stable.

(b) If at least one eigenvalue of $DH(\bar{X})$ has absolute value greater than one, then the equilibrium $\bar{X}$ of Eq.(1.3.15) is unstable.

In general, in this book, we are interested in scalar difference equations of the form of Eq.(1.3.1) where the function $F(u_0, u_1, \ldots, u_k)$ has continuous partial derivatives. Then the linearized equation associated with Eq.(1.3.1) about an equilibrium point $\bar{x}$ is

$$y_{n+1} = \sum_{i=0}^{k} \frac{\partial F}{\partial u_i}(\bar{x}, \ldots, \bar{x}) y_{n-i}, \ n = 0, 1, \ldots.$$

Corollary 1.3.2 *Assume that F is a $\mathbf{C^1}$ function and let $\bar{x}$ be an equilibrium of Eq.(1.3.1). Then the following statements are true:*

(a) If all the roots of the polynomial equation

$$\lambda^{k+1} - \sum_{i=0}^{k} \frac{\partial F}{\partial u_i}(\bar{x}, \ldots, \bar{x}) \lambda^{k-i} = 0 \tag{1.3.16}$$

lie in the open unit disk $|\lambda| < 1$, then the equilibrium $\bar{x}$ of Eq.(1.3.1) is asymptotically stable.

(b) If at least one root of Eq.(1.3.16) has absolute value greater than one, then the equilibrium $\bar{x}$ of Eq.(1.3.1) is unstable.

Let $\bar{X}$ be an equilibrium of Eq.(1.3.15) and assume that H is a $\mathbf{C^1}$ function in an open neighborhood of $\bar{X}$. Let $DH(\bar{X})$ denote the Jacobian matrix of the function H at $\bar{X}$. Then the following terminology is often used in the literature.

$\bar{X}$ is called a **hyperbolic equilibrium** if $DH(\bar{X})$ has no eigenvalues with absolute value equal to 1.

$\bar{X}$ is called a **sink** or an **attracting equilibrium** if every eigenvalue of $DH(\bar{X})$ has absolute value less than 1.

$\bar{X}$ is called a **source** or a **repelling equilibrium** if every eigenvalue of $DH(\bar{X})$ has absolute value greater than 1.

$\bar{X}$ is called a **saddle point** if some of the eigenvalues of $DH(\bar{X})$ are larger and some are less than 1 in absolute value.

When one of the eigenvalues of the Jacobian matrix $DH(\bar{X})$ has absolute value equal to one, the stability character of the equilibrium $\bar{X}$ of Eq.(1.3.15) is very difficult to determine and requires, in general, special considerations. Before we can state an important result in this direction, when H is a scalar function, we need the following definition.

Definition 1.3.3 *Schwarzian Derivative*

*Let f be a $\mathbf{C}^3$ function in some interval I. The **Schwarzian** derivative $Sf(x)$ of f at a point $x \in I$, where $f'(x) \neq 0$, is given by*

$$Sf(x) = \frac{f'''(x)}{f'(x)} - \frac{3}{2}\left(\frac{f''(x)}{f'(x)}\right)^2. \tag{1.3.17}$$

See Singer [1] and Sandefur [1].

Theorem 1.3.8 *Let $\bar{x}$ be an equilibrium point of the scalar difference equation*

$$x_{n+1} = f(x_n), \ n = 0, 1, \ldots$$

and assume that f is a $\mathbf{C}^3$ function such that

$$f'(\bar{x}) = -1.$$

Then the following statements are true:

(a) If $Sf(\bar{x}) < 0$, then $\bar{x}$ is an attracting equilibrium.

(b) If $Sf(\bar{x}) > 0$, then $\bar{x}$ is a repelling equilibrium.

1.4   Periodicity and General Background

Throughout this section p will denote a positive integer.

Definition 1.4.1 *Periodicity*

*(a) A sequence $\{x_n\}$ is said to be **periodic with period** p if*

$$x_{n+p} = x_n \ \text{ for } \ n = 0, 1, \ldots. \tag{1.4.1}$$

*(b) A sequence $\{x_n\}$ is said to be **periodic with prime period** p, or with **minimal period** p, if it is periodic with period p and p is the least positive integer for which (1.4.1) holds.*

A difference equation is called a p–**cycle**, if every solution of it is periodic with period p. For example Eq.(1.1.8) is a 5–cycle and Eq.(1.1.9) is an 8–cycle.

Suppose that $\{X_n\}$ is a periodic solution with period p of the difference equation

$$X_{n+1} = H(X_n), \ n = 0, 1, \ldots \tag{1.4.2}$$

where $X_n \in \mathbf{R}^{k+1}$ for every $n \geq 0$ and $H \in \mathbf{C}^1[\mathbf{R}^{k+1}, \mathbf{R}^{k+1}]$. Then one can see that each of the p points $X_0, \ldots, X_{p-1}$ is an equilibrium of the difference equation

$$Z_{n+1} = H^p(Z_n), \ n = 0, 1, \ldots$$

where H^p is the pth-iterate of H. Also, by the chain rule one can see that the Jacobian matrix of H^p at each of the p points $X_0, \ldots, X_{p-1}$ has the same eigenvalues. When the function H in Eq.(1.4.2) is such that its pth-iterate H^p is a $\mathbf{C}^1$ function, then the following theorem states that the stability of the periodic solution $\{X_n\}$ is determined by the eigenvalues of the Jacobian matrix $DH^p(X_0)$ of the function H^p evaluated at X_0. For a proof see, for example, Devaney [1].

Theorem 1.4.1 *Let $\{X_n\}$ be a periodic solution with period p of Eq. (1.4.2) and assume that the pth-iterate H^p of the function H is a $\mathbf{C}^1$ function. Then the following statements are true:*

(a) *If all the eigenvalues of the Jacobian matrix $DH^p(X_0)$ lie in the open unit disk $|\lambda| < 1$, then the periodic solution $\{X_n\}$ of Eq.(1.4.2) is asymptotically stable.*

(b) *If at least one eigenvalue of $DH^p(X_0)$ has absolute value greater than one, then the periodic solution $\{X_n\}$ of Eq.(1.4.2) is unstable.*

The following theorem of Kurshan and Gopinath [1] is a kind of linearized periodicity result for nonlinear difference equations. This result gives necessary conditions for all solutions of a nonlinear difference equation to be periodic with the same period.

Theorem 1.4.2 *Let A be a constant $k \times k$ matrix and let $U \subset \mathbf{R}^k$ be an open set containing $0 \in \mathbf{R}^k$. Consider the nonlinear difference equation*

$$X_{n+1} = AX_n + F(X_n), \ n = 0, 1, \ldots \tag{1.4.3}$$

where $X_n \in \mathbf{R}^k$ for every $n \geq 0$ and $F : U \to U$ is an analytic function. Suppose that (1.3.12) is satisfied and that all solutions of Eq.(1.4.3) with $X_0 \in U$ are periodic with period p. Then the following statements are true:

(a) *all solutions of the linear equation*

$$X_{n+1} = AX_n, \ n = 0, 1, \ldots \tag{1.4.4}$$

are periodic with period p.

(b) $A^p = I$, *where I is the identity matrix.*

(c) *All eigenvalues of A are distinct and for every eigenvalue λ of A, $\lambda^p = 1$.*

Remark 1.4.1 In Theorem 1.4.2 the assumption that the function F is analytic is essential. For the convenience of the reader we present below the definition of analytic function.

Let $U \subset \mathbf{R}^k$ be a neighborhood of a point $X_0 \in \mathbf{R}^k$. Let $\mathbf{N}$ be the set of nonnegative integers and let $\alpha = [\alpha_1, \ldots, \alpha_k]^T \in \mathbf{N}^k$. For a vector $X = [x_1, \ldots, x_k]^T$, we introduce the notation X^α to denote the product $x^{\alpha_1} \ldots x_k^{\alpha_k}$. We say that a scalar valued function g that maps U into $\mathbf{R}$ is **analytic** in the neighborhood U of X_0 if there are scalars $c(\alpha)$ $(\alpha \in \mathbf{N}^k)$ such that for each vector $X \in U$ and each 1-1 and onto map $\omega : \mathbf{N} \to \mathbf{N}^k$, the series

$$\sum_{i=0}^{\infty} c(\omega(i))(X - X_0)^{\omega(i)}$$

converges to $g(X)$.

A vector–valued function $F = [f_1, \ldots, f_k]^T$ which maps U into $\mathbf{R}^k$ is said to be analytic in U if each component f_i, for $i = 1, \ldots, k$, is analytic in U.

1.5 Global Attractivity for First-Order Equations

In this section we state some known global attractivity results about first-order equations which will be useful in the sequel.

In a number of papers, Cull [1-3], Huang [1-4], and Rosenkranz [1] have investigated the problem of the global attractivity of a special kind of first-order difference equation known as "population model". Before we state the results we need the following definition.

Definition 1.5.1 *Population Model*
Let $A \in (0, \infty]$ and let $g : [0, A) \to [0, A)$ be a continuous function. Suppose $g(0) = 0$, $g(x) > 0$ for $0 < x < A$, and assume that g has a unique fixed point $\bar{x} \in (0, A)$. Suppose also that $g(x) > x$ for $0 < x < \bar{x}$ and $g(x) < x$ for $\bar{x} < x < A$. Then the difference equation

$$x_{n+1} = g(x_n), \quad n = 0, 1, \ldots \tag{1.5.1}$$

*is called a **population model**.*

The following theorem from Huang [3] was originally proved with stronger hypotheses in Cull [1], (see also Cull [2,3] and Rosenkranz [1]), and gives sufficient and necessary and sufficient conditions for global attractivity in population models.

Theorem 1.5.1 *Let $A \in (0, \infty]$ and assume that Eq.(1.5.1) is a population model as described by definition 1.3.1. Then the following statements are true:*

(a) If $g(x) \leq \bar{x}$ for $x < \bar{x}$, then $\bar{x}$ is a global attractor of all solutions of Eq.(1.5.1) with $x_0 \in (0, A)$.

(b) Let x_m be the rightmost of all absolute maximum points of $g(x)$ in $(0, \bar{x}]$ and assume that $g(x_m) > \bar{x}$. Then $\bar{x}$ is a global attractor of all solutions of Eq.(1.5.1) with $x_0 \in (0, A)$ if and only if, $g(g(x)) > x$ for all $x \in [x_m, \bar{x})$.

The following result from Cull [3] gives necessary and sufficient conditions for global attractivity in population models.

Theorem 1.5.2 *Let $A \in (0, \infty]$ and assume that Eq.(1.5.1) is a population model as described by definition 1.3.1. Suppose that one of the following two hypotheses is satisfied.*

(H_1) *g has no maximum in $(0, \bar{x})$.*

(H_2) *There is a maximum of g at a point $x_m \in (0, \bar{x})$ for which g is strictly monotone decreasing for $x > x_m$.*

Then $\bar{x}$ is a global attractor of all positive solutions of Eq.(1.5.1) with $x_0 \in (0, A)$ if and only if g has no periodic points of prime period 2. Furthermore when (H_1) is satisfied, then $\{x_n\}$ converges eventually monotonically to $\bar{x}$.

The next result due to Fisher, Goh and Vincent [1], requires the existence of a "Liapunov function" V.

Theorem 1.5.3 *Let $A \in (0, \infty]$ and let $f \in C[(0, A), (0, A)]$. Assume that the following conditions are satisfied:*

(a) *The function f has a unique fixed point $\bar{x} \in (0, A)$.*

(b) *There exists a continuous function $V : (0, A) \to [0, \infty)$ such that the following three conditions are satisfied:*

    (i) *V is nonincreasing in a neighborhood of 0;*

    (ii) *V is nondecreasing in a neighborhood of A;*

    (iii) *$V(f(x)) < f(x)$ for all $x \in (0, A)$ with $x \neq \bar{x}$.*

Then $\bar{x}$ is a global attractor of all solutions of the difference equation

$$x_{n+1} = f(x_n)$$

with $x_0 \in (0, A)$.

1.6   Some Basic Lemmas

In this section we present some basic lemmas which will be useful in the sequel. The proofs of the first two lemmas are by simple induction and will be omitted.

Lemma 1.6.1 *Let $\alpha > 0$ and $\beta \in \mathbf{R}$ and assume that $\{x_n\}$, $\{y_n\}$, and $\{z_n\}$ are sequences of real numbers such that $x_0 \leq y_0 \leq z_0$ and*

$$\left.\begin{array}{rcl} x_{n+1} & \leq & \alpha x_n + \beta \\[2mm] y_{n+1} & = & \alpha y_n + \beta \\[2mm] z_{n+1} & \geq & \alpha z_n + \beta \end{array}\right\}, \; n = 0, 1, \ldots .$$

Then

$$x_n \leq y_n \leq z_n, \; n = 0, 1, \ldots .$$

Lemma 1.6.2 *Let $\alpha, \beta \in \mathbf{R}$ and let y_0 be given. Then the unique solution of the equation*

$$y_{n+1} = \alpha y_n + \beta, \; n = 0, 1, \ldots$$

is

$$y_n = \begin{cases} y_0 + n\beta & \text{if } \alpha = 1 \\[3mm] y_0 \alpha^n + \dfrac{\beta}{1 - \alpha}(1 - \alpha^n) & \text{if } \alpha \neq 1. \end{cases}$$

Lemma 1.6.3 *Let $F \in C[[0, \infty), (0, \infty)]$ be a nonincreasing function. and let $\bar{x}$ denote the (unique) fixed point of F. Then the following statements are equivalent:*

(a) $\bar{x}$ is the only fixed point of F^2 in $(0, \infty)$;

(b) $F^2(x) > x$ for $0 < x < \bar{x}$;

(c) If λ and μ are positive numbers such that

$$F(\mu) \leq \lambda \leq \bar{x} \leq \mu \leq F(\lambda), \tag{1.6.1}$$

then

$$\lambda = \mu = \bar{x}. \tag{1.6.2}$$

(d) $\bar{x}$ is a global attractor of all positive solutions of the equation

$$x_{n+1} = F(x_n), \; n = 0, 1, \ldots \tag{1.6.3}$$

with $x_0 \in [0, \infty)$.

Proof. (a) $\Rightarrow$ (b). Otherwise there exists an $x \in (0, \bar{x})$, such that $F^2(x) \leq x$. But $F^2(0) > 0$ and so F^2 has fixed points in $(0, \bar{x})$ which is impossible.

(b) $\Rightarrow$(c). Let λ and μ be positive numbers such that (1.6.1) holds. We now claim that $\lambda = \bar{x}$. Otherwise $\lambda < \bar{x} \leq \mu$ and so $\lambda \geq F(\mu) \geq F^2(\lambda) > \lambda$ which is impossible. Similarly we can see that $\mu = \bar{x}$ and so (1.6.2) holds.

(c) $\Rightarrow$(a). Otherwise there exists an $x_0 \in (0, \bar{x}) \cup (\bar{x}, \infty)$ such that $F^2(x_0) = x_0$. If $x_0 \in (0, \bar{x})$, by taking $\lambda = x_0$ and $\mu = F(x_0)$, we can see that (1.6.1) holds but not (1.6.2). On the other hand if $x_0 \in (\bar{x}, \infty)$, by taking $\lambda = F(x_0)$ and $\mu = x_0$, we see that again (1.6.1) holds but not (1.6.2). The proof is complete.

(d) $\Rightarrow$(a). Assume, for the sake of contradiction, that $x' \neq \bar{x}$ is a fixed point of F^2 on $(0, \infty)$. Then the solution $\{x_n\}$ of Eq.(1.6.3) with $x_0 = x'$ has the property that

$$\lim_{n \to \infty} x_{2n} = x' \neq \bar{x},$$

which is a contradiction since $\bar{x}$ is a global attractor of all positive solutions of Eq.(1.6.3).

(a) $\Rightarrow$(d). If $x_0 = \bar{x}$, then $x_n = \bar{x}$ for all $n \geq 0$ and (d) is proved. Next, assume that $0 < x_0 < \bar{x}$. Since the function F^2 is nondecreasing, it follows by induction that $0 < x_{2n} < \bar{x}$. Furthermore from (b), which is equivalent to (a), we obtain

$$x_{2n} = F(x_{2n-1}) = F^2(x_{2n-2}) > x_{2n-2}$$

and the subsequence $\{x_{2n}\}$ is convergent. Let

$$x = \lim_{n \to \infty} x_{2n}.$$

Then $x = F^2(x)$, and from (a) we see that $x = \bar{x}$. Since $x_{2n+1} = F(x_{2n})$, it follows that the subsequence $\{x_{2n+1}\}$ is also convergent and we have

$$\lim_{n \to \infty} x_{2n+1} = \lim_{n \to \infty} F(x_{2n}) = F(\bar{x}) = \bar{x}.$$

To complete the proof it remains to consider the case where $\bar{x} < x_0$. Since F is nonincreasing we obtain $x_1 = F(x_0) \leq F(\bar{x}) = \bar{x}$. Then by a procedure similar to that which was given above, we obtain

$$\lim_{n \to \infty} x_n = \bar{x}.$$

The proof is complete.   □

Lemma 1.6.4 *Let $F, H \in \mathbf{C}[[0, \infty), (0, \infty)]$ be nonincreasing in $[0, \infty)$ and let $\bar{x} \in (0, \infty)$ be such that*

$$F(\bar{x}) = H(\bar{x}) = \bar{x}$$

and

$$[H(x) - F(x)](x - \bar{x}) \leq 0 \quad \text{for} \quad x \geq 0. \tag{1.6.4}$$

Assume that $\bar{x}$ is the only fixed point of H^2 in $(0, \infty)$. Then $\bar{x}$ is also the only fixed point of F^2 in $(0, \infty)$.

Proof. Otherwise by Lemma 1.6.3, there exist positive numbers λ and μ such that (1.6.1) holds but not (1.6.2). Then in view of (1.6.1) and (1.6.4),

$$H(\mu) \leq F(\mu) \leq \lambda \leq \bar{x} \leq \mu \leq F(\lambda) \leq H(\lambda)$$

which by Lemma 1.6.3 again implies that $\lambda = \mu$. This is a contradiction and the proof is complete. □

Lemma 1.6.5 *Let α be a positive number and consider the difference equation*

$$x_{n+1} = f(x_n), \quad n = 0, 1, \ldots \tag{1.6.5}$$

where $f \in C[(0, \infty), (\alpha, \infty)]$. Assume that $f(u)$ is nonincreasing in u and let $\bar{x}$ denote the unique fixed point of f. Assume that the only solution of the system

$$U = f(L) \quad \text{and} \quad L = f(U) \tag{1.6.6}$$

is $U = L = \bar{x}$. Then the following statements are true:

(a) $\bar{x}$ is a global attractor of all positive solutions of Eq.(1.6.5).

(b) The subsequences $\{x_{2n}\}$ and $\{x_{2n+1}\}$ of every positive solution of Eq.(1.6.5) converge monotonically to $\bar{x}$.

(c) Every positive solution of Eq.(1.6.5) oscillates about $\bar{x}$ and the length of each nontrivial semicycle is one; that is,

$$(x_{n+1} - \bar{x})(x_n - \bar{x}) < 0, \quad n = 0, 1, \ldots.$$

Proof. Let $\{x_n\}$ be a positive solution of Eq.(1.6.5). We must prove that

$$\lim_{n \to \infty} x_n = \bar{x}$$

and that statements (b) and (c) hold. The proof is obvious if some iterate of x_0 is equal to $\bar{x}$. We will give the proof when $x_0 < \bar{x}$ and under the assumption that no iterate of x_0 is equal to $\bar{x}$. The case where $x_0 > \bar{x}$ is similar and will be omitted. Then

$$x_1 = f(x_0) > f(\bar{x}) = \bar{x} \quad \text{and} \quad x_2 = f(x_1) < f(\bar{x}) = \bar{x}$$

and by induction

$$x_{2n} < \bar{x} < x_{2n+1} \quad \text{for} \quad n = 0, 1, \ldots$$

and (c) is proved.

Looking at x_2 we see there are three possibilities:

$$x_2 = x_0, \quad x_2 < x_0, \quad \text{or} \quad x_2 > x_0.$$

If $x_2 = x_0$, then $x_3 = x_1$ and $\{x_0, x_1\}$ is a solution of (1.6.6). This contradicts the fact that the only solution of (1.6.6) is $\{\bar{x}, \bar{x}\}$. Hence $x_2 \neq x_0$.

Next consider the case $x_2 < x_0$. This implies that

$$x_3 = f(x_2) > f(x_0) = x_1$$

and so by induction

$$0 < \alpha < x_{2n+2} < x_{2n} < \bar{x} < x_{2n+1} < x_{2n+3} < f(\alpha) \quad \text{for} \quad n = 0, 1, \ldots.$$

Thus there exist positive numbers $\bar{L}$ and $\bar{U}$ such that

$$\lim_{n \to \infty} \{x_{2n}\} = \bar{L} \quad \text{and} \quad \lim_{n \to \infty} \{x_{2n+1}\} = \bar{U}.$$

It follows from (1.6.5) that

$$\bar{L} = f(\bar{U}) \quad \text{and} \quad \bar{U} = f(\bar{L})$$

and so $\{\bar{L}, \bar{U}\}$ is a solution of (1.6.6). This is impossible because $\bar{L} < \bar{x} < \bar{U}$, and so it must be true that $x_2 > x_0$. Thus

$$x_3 = f(x_2) < f(x_0) = x_1 \quad \text{and} \quad x_3 = f(x_2) > f(\bar{x}) = \bar{x}$$

and by induction

$$0 < \alpha < x_{2n} < x_{2n+2} < \bar{x} < x_{2n+3} < x_{2n+1}, \quad n = 0, 1, \ldots.$$

From this we conclude that (a) and (b) hold and the proof is complete. $\qquad\square$

Remark 1.6.1 The conclusion of Lemma 1.6.5, remains true if instead of assuming that $f \in C[(0, \infty), (\alpha, \infty)]$ for some $\alpha > 0$, we assume that $f \in C[[0, \infty), (0, \infty)]$.

The following technical lemma will be useful in Chapter 3.

Lemma 1.6.6 *Let $\alpha \in (1, \infty)$ and $\beta \in (0, \infty)$, and let p be a positive integer. Set $\gamma = (\alpha - 1)/\beta$ and define the function φ_p by*

$$\varphi_p(x) = \gamma(\frac{\alpha}{1 + \beta x})^p \quad \text{for} \quad x \geq 0. \tag{1.6.7}$$

Assume that

$$(\alpha - 1)(p - 1) \leq 1. \tag{1.6.8}$$

Then the only positive solution of the equation

$$\varphi_p(\varphi_p(x)) = x \tag{1.6.9}$$

is $x = \gamma$.

Proof. Set

$$\psi(x) = x - \varphi_p(\varphi_p(x)) = x - \alpha^p\left(\frac{\alpha - 1}{\beta}\right)\left[1 + \frac{(\alpha - 1)\alpha^p}{(1 + \beta x)^p}\right]^{-p}$$

and observe that $\psi(0) < 0$, $\psi(\gamma) = 0$, and $\psi(\infty) = \infty$. Clearly it suffices to show that

$$\psi'(x) > 0 \quad \text{for} \quad x \geq 0 \quad \text{and} \quad x \neq \gamma. \tag{1.6.10}$$

We have

$$\psi'(x) = 1 - (\alpha - 1)^2 p^2 \alpha^{2p}\left[1 + \beta x + \frac{(\alpha - 1)\alpha^p}{(1 + \beta x)^{p-1}}\right]^{-p-1}$$

and so $\psi'(x) > 0$ for $p = 1$. For $p \geq 2$, set

$$\eta(t) = t + (\alpha - 1)\alpha^p t^{-p+1}$$

and observe that η has a minimum in $(0, \infty)$ at $t_0 = \alpha((p - 1)(\alpha - 1))^{1/p}$. Hence

$$\eta(t) \geq \eta(t_0) = \frac{\alpha p((p - 1)(\alpha - 1))^{1/p}}{p - 1}$$

and so for $p \geq 2$,

$$\psi'(x) \geq 1 - [((p - 1)(\alpha - 1))^{1/p}\alpha(\frac{p - 1}{p})]^{p-1}. \tag{1.6.11}$$

In view of (1.6.8) we see that,

$$\alpha \frac{p - 1}{p} \leq 1$$

and so (1.6.11) implies that $\psi'(x) \geq 0$ for $x \geq 0$.

To complete the proof it remains to show that $\psi'(x) > 0$ for $x \neq \gamma$. Now $\psi'(x) = 0$ if and only if $\alpha = 1 + \frac{1}{p-1}$ and $1 + \beta x = t_0 = \alpha((p - 1)(\alpha - 1))^{1/p}$, from which it follows that $x = \gamma$ and the proof is complete. $\qquad\square$

1.7 Some Useful Theorems from Analysis

In this section we state for the convenience of the reader some results from analysis which are needed in some parts of this book.

Let X be a nonempty set. A **distance** on X is a function $d : X \times X \to [0, \infty)$ having the following properties for all $x, y, z \in X$:

(a) $d(x, y) = 0$ if and only if $x = y$;

(b) $d(x, y) = d(y, x)$;

(c) $d(x, z) \leq d(x, y) + d(y, z)$ (triangle inequality);

A **metric space** is a pair (X, d), where d is a distance on X. Let $\{x_n\}$ be a sequence of points in a metric space X. The sequence $\{x_n\}$ is **bounded** if there exists $a \in X$ and a positive constant B such that for all $n \geq 0$, $d(x_n, a) < B$. We say that $\{x_n\}$ **converges** to $x \in X$ if the sequence of nonnegative numbers $\{d(x_n, x)\}$ converges to zero. A sequence $\{x_n\}$ is called a **Cauchy sequence** if for every $\epsilon > 0$ there exists a positive integer $N = N(\epsilon)$ such that for all $n, m \geq N(\epsilon)$, $d(x_n, x_m) < \epsilon$. Clearly a convergent sequence is a Cauchy sequence, but the converse may not be true. However a Cauchy sequence is a bounded sequence. A **complete metric space** is a metric space (X, d) in which every Cauchy sequence converges to a point in X.

Let (X, d) be a metric space and let $T : X \to X$. We say that T is a **contraction mapping** on X if there exists a number $r \in [0, 1)$ such that

$$d(Tx, Ty) \leq rd(x, y) \quad \text{for every} \quad x, y \in X.$$

Theorem 1.7.1 *(The Banach Contraction Principle) Let (X, d) be a complete metric space and let T be a contraction mapping on X. Then T has exactly one fixed point on X. That is, there exists exactly one $\bar{x} \in X$ such that $T\bar{x} = \bar{x}$.*

A **norm** $\| \cdot \|$ on a vector space X is a function that maps X into $[0, \infty)$ having the following properties for all vectors $x, y \in X$ and all complex numbers α:

(a) $\|x\| = 0$ if and only if $x = 0$;

(b) $\|\alpha x\| = |\alpha| \|x\|$;

(c) $\|x + y\| \leq \|x\| + \|y\|$ (triangle inequality);

A vector space X on which we have defined a norm $\| \cdot \|$ is called a **normed vector space**. Clearly, a normed vector space X is also a metric space with the distance induced by the norm $d(x, y) = \|x - y\|$. A **Banach space** is a normed vector space which is also a complete metric space with the distance induced by the norm.

Example 1.7.1 An important example of a Banach space for our investigations is the space ℓ_∞ of all real bounded sequences $x = \{x_n\}$ with the norm

$$\|x\| = \sup_{n \geq 0} |x_n|.$$

Let M be a subset of a Banach space X. We say that a point $x \in X$ is a **limit point** of M if there exists a sequence of vectors in M which converges to x. We say that M is **closed** if M contains all of its limit points.

Theorem 1.7.2 *Let X be a complete metric space, and let M be a subspace of X. Then M is complete if and only if M is closed.*

The union of M and its limit points is called the **closure** of M and is denoted by $\overline{M}$.

A set M is called **convex** if for every $x, y \in M$ and for every $\lambda \in [0, 1]$,

$$\lambda x + (1 - \lambda)y \in M.$$

A subset M of a Banach space X is called **compact** if every sequence of vectors in M contains a subsequence which converges to a vector in M. We say that M is **relatively compact** if every sequence of vectors in M contains a subsequence which converges to a vector in X. A set M is **bounded** if there exists a positive constant B such that for all vectors $x \in M$, $\|x\| < B$.

1.8 Notes

A good source for the proofs of the oscillation results which we need in this book is Györi and Ladas [2]. The books Agarwal [1], Kelley and Peterson [1], and Lakshmikantham and Trigiante [1] give a good introduction to the basic topics on linear difference equations and on linearized stability.

Explicit criteria for all roots of Eq.(1.3.7) to lie in the open unit disk $|\lambda| < 1$ are also provided by the so-called Jury test. See Jury [1] and Lewis [1]. For some related work in this area see also Siljak [1-2].

In addition to the global attractivity results which we mentioned in Section 1.5, the reader should consult the paper of Singer [1] from where such results follow for functions with negative Schwarzian derivative.

In particular we mention the following results for the scalar difference equation

$$x_{n+1} = f(x_n), n = 0, 1, \ldots \tag{1.8.1}$$

where f maps some interval I into itself. We will assume that Eq.(1.8.1) has a unique equilibrium which we denote by $\bar{x}$.

Theorem 1.8.1 *Assume that $\bar{x}$ is locally asymptotically stable and let f has negative Schwarzian derivative everywhere on I (except for points x^*, where $f'(x^*) = 0$). Then in each of the following cases, $\bar{x}$ is globally asymptotically stable:*

(a) f is decreasing on I;

(b) f is unimodal on I, that is, there exists a point c in the interior of I such that f is increasing for $x < c$ and decreasing for $x > c$, or vice versa.

Lemmas 1.6.3, 1.6.4 and 1.6.6 are from Kocic and Ladas [5]. Lemma 1.6.5 is from DeVault, Kocic and Ladas [1].

Chapter 2

Global Stability Results

In this chapter we present some general results on the global asymptotic stability and global attractivity of some quite general nonlinear difference equations of order greater than one. The global asymptotic stability of such equations is an unexplored area, still in its infancy, with great potential for applications. Therefore this is a fertile area for research and, hopefully, the results in this chapter will stimulate interest for further investigations.

2.1   Global Asymptotic Stability of $x_{n+1} = x_n f(x_n, x_{n-1})$

Our aim in this section is to establish the following global asymptotic stability result for the nonlinear second-order difference equation

$$x_{n+1} = x_n f(x_n, x_{n-1}), \ n = 0, 1, \ldots. \tag{2.1.1}$$

For applications of this result see Sections 4.1, 4.2 and 6.1.

Theorem 2.1.1 *Assume that $f \in C[(0, \infty) \times (0, \infty), (0, \infty)]$, $f(u, v)$ is nonincreasing in u and decreasing in v, and $uf(u, u)$ is increasing in u. Suppose that Eq.(2.1.1) has a unique positive equilibrium $\bar{x}$. Then $\bar{x}$ is globally asymptotically stable.*

Proof. Let $\{x_n\}$ be a solution of Eq.(2.1.1) with initial conditions $x_{-1}, x_0 \in (0, \infty)$. We must prove that $\bar{x}$ is stable and that

$$\lim_{n \to \infty} x_n = \bar{x}. \tag{2.1.2}$$

The proof is long and will be accomplished by means of a thorough and detailed study of the positive and negative semicycles of the solution $\{x_n\}$.

A **positive semicycle** of $\{x_n\}$ consists of a "string" of terms $\{x_{l+1}, x_{l+2}, \ldots, x_m\}$, all greater than or equal to $\bar{x}$, with $l \geq -2$ and $m \leq \infty$ and such that

$$\text{either} \quad l = -2 \quad \text{or} \quad l \geq -1 \quad \text{and} \quad x_l < \bar{x}$$

and

$$\text{either} \quad m = \infty \quad \text{or} \quad m < \infty \quad \text{and} \quad x_{m+1} < \bar{x}.$$

A **negative semicycle** of $\{x_n\}$ consists of a "string" of terms $\{x_{k+1}, x_{k+2}, \ldots, x_l\}$, all less than $\bar{x}$, with $k \geq -2$ and $l \leq \infty$ and such that

$$\text{either} \quad k = -2 \quad \text{or} \quad k \geq -1 \quad \text{and} \quad x_k \geq \bar{x}$$

and

$$\text{either} \quad l = \infty \quad \text{or} \quad l < \infty \quad \text{and} \quad x_{l+1} \geq \bar{x}.$$

The first semicycle of a solution starts with the term x_{-1} and is positive if $x_{-1} \geq \bar{x}$ and negative if $x_{-1} < \bar{x}$.

A solution may have a finite number of semicycles or infinitely many.

The following technical lemma which is a consequence of the hypotheses imposed on f will be very useful.

Lemma 2.1.1 *Assume that $0 < a < \bar{x} < b$. Then the following inequalities are true:*

$$af(a, a) < \bar{x}; \quad bf(b, b) > \bar{x}. \tag{2.1.3}$$

$$\bar{x} f\left(\frac{\bar{x}^2}{a}, \frac{\bar{x}^2}{a}\right) > a; \quad \bar{x} f\left(\frac{\bar{x}^2}{b}, \frac{\bar{x}^2}{b}\right) < b. \tag{2.1.4}$$

$$\bar{x} f(\bar{x} f(a, a), \bar{x} f(a, a)) > a; \quad \bar{x} f(\bar{x} f(b, b), \bar{x} f(b, b)) < b. \tag{2.1.5}$$

Proof. Inequalities (2.1.3) are a simple consequence of the increasing character of the function $u f(u, u)$ and the fact that $f(\bar{x}, \bar{x}) = 1$.

Since $\bar{x}^2/a > \bar{x}$, the second inequality in (2.1.3) yields

$$\frac{\bar{x}^2}{a} f\left(\frac{\bar{x}^2}{a}, \frac{\bar{x}^2}{a}\right) > \bar{x}$$

which proves the first inequality in (2.1.4). The second inequality in (2.1.4) is proved in a similar way.

From (2.1.3), $\bar{x} f(a, a) < \bar{x}^2/a$, and so from (2.1.4) and the fact that $f(u, u)$ decreases in u, we have $\bar{x} f(\bar{x} f(a, a), \bar{x} f(a, a)) > a$. The second inequality in (2.1.5) is proved similarly. $\square$

The next lemma presents a detailed description of the semicycles of $\{x_n\}$.

Lemma 2.1.2 *The following statements are true:*

(a) If x_{-1} and x_0 are not both equal to $\bar{x}$, a positive semicycle cannot have two consecutive terms equal to $\bar{x}$.

(b) Every semicycle of $\{x_n\}$, except perhaps for the first one, has at least two terms.

(c) The extreme in a semicycle is equal to the first or to the second term of the semicycle. Furthermore after the first term, the remaining terms in a positive semicycle are nonincreasing and the remaining terms in a negative semicycle are nondecreasing.

(d) Except perhaps for the first semicycle of a solution, in a semicycle with finitely many terms, the extreme of the semicycle cannot be equal to the last term.

(e) In a solution with four or more semicycles, the maxima in successive positive semicycles are decreasing and the minima in successive negative semicycles are increasing.

Proof. The proof of (a) is simple and will be omitted. For the remaining statements we will only give the proof for positive semicycles whose terms are not all equal to $\bar{x}$. The proof for negative semicycles is similar and will be omitted. The proof for the trivial semicycle where $x_{-1} = x_0 = \bar{x}$ is obvious.

(b) If x_n is the first term in a positive semicycle (other than the first semicycle), then

$$x_{n+1} = x_n f(x_n, x_{n-1}) > x_n f(x_n, x_n) \geq \bar{x} f(\bar{x}, \bar{x}) = \bar{x}$$

so x_{n+1} is also in the same semicycle.

(c) If x_n, x_{n+1} are two consecutive terms in a positive semicycle, then

$$x_{n+2} = x_{n+1} f(x_{n+1}, x_n) \leq x_{n+1} f(\bar{x}, \bar{x}) = x_{n+1}. \tag{2.1.6}$$

(d) Assume for the sake of contradiction that the maximum in a positive semicycle is equal to the last term. Let x_n and x_{n+1} be the last two terms in this positive semicycle. Then

$$x_n \leq x_{n+1} \quad \text{and} \quad x_{n+2} < \bar{x}.$$

On the other hand, by using the monotonic character of $f(u, v)$ and $uf(u, u)$ we obtain

$$x_{n+2} = x_{n+1} f(x_n, x_{n+1}) \geq x_{n+1} f(x_{n+1}, x_{n+1}) > \bar{x} f(\bar{x}, \bar{x}) = \bar{x}$$

which is a contradiction.

(e) Consider four consecutive semicycles

$$C_{r-1} \;=\; \{x_{k+1}, x_{k+2}, \ldots, x_l\} \quad \text{— negative semicycle,}$$

$$C_r \;=\; \{x_{l+1}, x_{l+2}, \ldots, x_m\} \quad \text{— positive semicycle,}$$

$$C_{r+1} \;=\; \{x_{m+1}, x_{m+2}, \ldots, x_n\} \quad \text{— negative semicycle,}$$

$$C_{r+2} \;=\; \{x_{n+1}, x_{n+2}, \ldots, x_p\} \quad \text{— positive semicycle.}$$

If b_{r-1}, b_r, b_{r+1} and b_{r+2} denote the extreme values in these four semicycles, respectively, we must prove that

$$b_{r+2} < b_r \quad \text{and} \quad b_{r-1} < b_{r+1}. \tag{2.1.7}$$

It follows from (c) that either $b_r = x_{l+1}$ or that $b_r = x_{l+2}$. In the first case

$$b_r = x_l f(x_l, x_{l-1}) < \bar{x} f(b_{r-1}, b_{r-1}).$$

In the second case

$$b_r \;=\; x_{l+1} f(x_{l+1}, x_l) = x_l f(x_l, x_{l-1}) f(x_{l+1}, x_l)$$

$$\leq\; x_l f(x_l, x_l) f(x_l, x_{l-1}) < \bar{x} f(b_{r-1}, b_{r-1}).$$

Thus in either case

$$b_r < \bar{x} f(b_{r-1}, b_{r-1}). \tag{2.1.8}$$

In a similar way we obtain

$$b_{r+1} \geq \bar{x} f(b_r, b_r). \tag{2.1.9}$$

Hence

$$b_{r+2} < \bar{x} f(b_{r+1}, b_{r+1}) < \bar{x} f(\bar{x} f(b_r, b_r), \bar{x} f(b_r, b_r)) \tag{2.1.10}$$

and

$$b_{r+1} > \bar{x} f(b_r, b_r) \geq \bar{x} f(\bar{x} f(b_{r-1}, b_{r-1}), \bar{x} f(b_{r-1}, b_{r-1})). \tag{2.1.11}$$

From these and (2.1.5), it follows that (2.1.7) holds and the proof of Lemma 2.1.2 is complete.   □

We are now ready to establish (2.1.2). This is a simple consequence of Lemma 2.1.2(c) if the solution has a finite number of semicycles. So assume that $\{x_n\}$ has infinitely many semicycles and set

$$\lambda = \liminf_{n \to \infty} x_n \quad \text{and} \quad \Lambda = \limsup_{n \to \infty} x_n$$

which by Lemma 2.1.2(e) both exist and satisfy

$$0 < \lambda \leq \bar{x} \leq \Lambda < \infty.$$

It follows from (2.1.10) and (2.1.11) that

$$\Lambda \leq \bar{x}f(\bar{x}f(\Lambda,\Lambda),\bar{x}f(\Lambda,\Lambda)) \quad \text{and} \quad \lambda \geq \bar{x}f(\bar{x}f(\lambda,\lambda),\bar{x}f(\lambda,\lambda)).$$

In view of (2.1.5), these imply that $\lambda = \bar{x} = \Lambda$ and (2.1.2) holds.

To complete the proof of the theorem it remains to prove that $\bar{x}$ is stable; that is, for every $\epsilon > 0$ there exists $\delta > 0$ such that

$$|x_{-1} - \bar{x}| < \delta \quad \text{and} \quad |x_0 - \bar{x}| < \delta \Rightarrow |x_n - \bar{x}| < \epsilon \quad \text{for} \quad n = 1, 2, \ldots . \qquad (2.1.12)$$

The following Lemma will enable us to establish local stability without the need of assuming differentiability.

Lemma 2.1.3 *Let* $\delta \in (0, \bar{x})$ *be given and assume that the initial conditions* x_{-1} *and* x_0 *are such that*

$$|x_{-1} - \bar{x}| < \delta \quad \text{and} \quad |x_0 - \bar{x}| < \delta. \qquad (2.1.13)$$

Then

$$\frac{-2\bar{x}\delta}{\bar{x} + \delta} < x_n - \bar{x} < \frac{2\bar{x}\delta}{\bar{x} - \delta} \quad \text{for} \quad n = 1, 2, \ldots . \qquad (2.1.14)$$

Proof. In view of (2.1.13), (2.1.14) is true for $n = -1$ and $n = 0$. Next, we will show that (2.1.14) is true for $n = 1$. To this end observe that in view of (2.1.13) and (2.1.3) we have

$$\begin{aligned}
x_1 &= x_0 f(x_0, x_{-1}) > (\bar{x} - \delta)f(\bar{x} + \delta, \bar{x} + \delta) \\
&= \frac{(\bar{x} - \delta)}{\bar{x} + \delta}(\bar{x} + \delta)f(\bar{x} + \delta, \bar{x} + \delta) > \frac{(\bar{x} - \delta)}{\bar{x} + \delta}\bar{x} \\
&= \bar{x} - \frac{2\bar{x}\delta}{\bar{x} + \delta}.
\end{aligned}$$

Similarly,

$$\begin{aligned}
x_1 &= x_0 f(x_0, x_{-1}) < (\bar{x} + \delta)f(\bar{x} - \delta, \bar{x} - \delta) \\
&= \frac{(\bar{x} + \delta)}{\bar{x} - \delta}(\bar{x} - \delta)f(\bar{x} - \delta, \bar{x} - \delta) < \frac{(\bar{x} + \delta)}{\bar{x} - \delta}\bar{x} \\
&= \bar{x} + \frac{2\bar{x}\delta}{\bar{x} - \delta}.
\end{aligned}$$

Therefore (2.1.14) is true for $n = 1$. From (2.1.13) and Lemma 2.1.1 (c), it is now clear that (2.1.14) is true if x_0 is contained in the last semicycle of the solution. Also, it is clear that the semicycle which contains x_0 contains at least two points. Now assume that x_0 is not contained in the last semicycle of the solution and denote by C_0 the semicycle which contains x_0 and by C_1 the next semicycle. For simplicity we will also assume that C_0 is a negative semicycle. The case when C_0 is a positive semicycle is similar and will be omitted. Let us denote by x^- and x^+ the minimum and the maximum, respectively, in the semicycles C_0 and C_1. Then by Lemma 2.1.2 it follows that

$$x^- \leq x_n \leq x^+ \quad \text{for} \quad n = 0, 1, \ldots$$

and so it suffices to prove that

$$\bar{x} - \frac{2\bar{x}\delta}{\bar{x} + \delta} < x^- < x^+ < \frac{2\bar{x}\delta}{\bar{x} - \delta} + \bar{x}. \tag{2.1.15}$$

The following two cases are now possible:

$$\text{(i)} \quad x_{-1} \in C_0; \tag{2.1.16}$$

$$\text{(ii)} \quad x_{-1} \notin C_0 \text{ and } x_1 \in C_0. \tag{2.1.17}$$

When (2.1.16) holds, then

$$\text{either} \quad x^- = x_{-1} \quad \text{or} \quad x^- = x_0$$

and because of (2.1.13)

$$\bar{x} - \frac{2\bar{x}\delta}{\bar{x} + \delta} < \bar{x} - \delta < x^- < \bar{x}. \tag{2.1.18}$$

Now in view of (2.1.8) and by using the monotonic properties of $f(u, u)$ and $uf(u, u)$ we have

$$\begin{aligned}
x^+ \; &< \; \bar{x}f(x^-, x^-) < \bar{x}f(\bar{x} - \delta, \bar{x} - \delta) \\[2mm]
&= \; \frac{\bar{x}}{\bar{x} - \delta}(\bar{x} - \delta)f(\bar{x} - \delta, \bar{x} - \delta) < \frac{\bar{x}}{\bar{x} - \delta}\bar{x}f(\bar{x}, \bar{x}) \\[2mm]
&= \; \frac{\bar{x}^2}{\bar{x} - \delta} < \bar{x} + \frac{2\bar{x}\delta}{\bar{x} - \delta}
\end{aligned} \tag{2.1.19}$$

and the proof of (2.1.15) is complete.

Next assume that (2.1.17) holds. Then $x_{-1} \geq \bar{x}$ and

$$\text{either} \quad x^- = x_0 \quad \text{or} \quad x^- = x_1.$$

If $x^- = x_0$, then the proof is identical to the one given above. On the other hand if $x^- = x_1$, then by using (2.1.13) and the second inequality in (2.1.3) we find

$$x^- = x_0 f(x_0, x_{-1}) > (\bar{x} - \delta) f(\bar{x} + \delta, \bar{x} + \delta)$$

$$> \frac{\bar{x} - \delta}{\bar{x} + \delta} \bar{x} = \bar{x} - \frac{2\bar{x}\delta}{\bar{x} + \delta}$$

which proves the first inequality in (2.1.15). The proof of the second inequality in (2.1.15) is similar to the one given above and will be omitted. □

We are now ready to finish the proof of Theorem 2.1.1. To this end, let $\epsilon > 0$ be given. Choose

$$\delta = \frac{\bar{x}\epsilon}{2\bar{x} + \epsilon}.$$

Then by Lemma 2.1.3, $|x_{-1} - \bar{x}| < \delta$ and $|x_0 - \bar{x}| < \delta$ imply $|x_n - \bar{x}| < \epsilon$ for $n \geq -1$. The proof is complete. □

The following corollary of Theorem 2.1.1 applies in the special case where Eq. (2.1.1) is of the form

$$x_{n+1} = x_n g(x_{n-1}), \quad n = 0, 1, \ldots \tag{2.1.20}$$

Corollary 2.1.1 *Assume that $g \in C[(0, \infty), (0, \infty)]$, $g(u)$ is decreasing in u, and $ug(u)$ is increasing in u. Let $\bar{x}$ denote the unique positive equilibrium of Eq.(2.1.20). Then $\bar{x}$ is globally asymptotically stable.*

In this case in addition to the statements (a) - (e) of Lemma 2.1.2, we have the following:

Lemma 2.1.4 *The semicycles of every positive solution, except perhaps for the first one, have at least three points each and the extreme occurs in the second term.*

Proof. We will give the proof for a positive semicycle. The proof for a negative semicycle is similar and will be omitted. To this end let x_n and x_{n+1} be the first two terms in a positive semicycle. Then

$$x_{n-1} < \bar{x} \leq x_n \quad \text{and} \quad x_{n+1} \geq \bar{x}.$$

We will now prove that

$$x_{n+1} > x_n \quad \text{and} \quad x_{n+2} > \bar{x}.$$

(which is more than we claimed). Indeed

$$x_{n+1} = x_n g(x_{n-1}) > x_n$$

and

$$x_{n+2} = x_{n+1}g(x_n) > x_n g(x_n) \geq \bar{x}g(\bar{x}) = \bar{x}$$

and the proof is complete.   □

The specific example which motivated the development of Theorem 2.1.1 is the discrete delay logistic model of Pielou [1,2]

$$x_{n+1} = \frac{\alpha x_n}{1 + \beta x_{n-1}}, \quad n = 0, 1, \dots \tag{2.1.21}$$

where

$$\alpha > 1 \quad \text{and} \quad \beta > 0 \tag{2.1.22}$$

and where $x_{-1} \in [0, \infty)$ and $x_0 \in (0, \infty)$.

For this equation we have the following consequence of Corollary 2.1.1.

Corollary 2.1.2 *Assume that (2.1.22) holds. Then the positive equilibrium $\bar{x} = (\alpha - 1)/\beta$ of Eq.(2.1.21) is globally asymptotically stable.*

For more details on this application, see Sections 3.4, 4.1.

Another example to which Theorem 2.1.1 applies is the rational recursive sequence

$$x_{n+1} = \frac{a + bx_n}{A + x_{n-1}}, \quad n = 0, 1, \dots \tag{2.1.23}$$

where $a, b, A \in (0, \infty)$ with $bA > a$ and $x_{-1}, x_0 \in [0, \infty)$. Here

$$f(u, v) = (\frac{a}{u} + b)\frac{1}{A + v} \tag{2.1.24}$$

and clearly, all the hypotheses of Theorem 2.1.1 are satisfied. Hence we have the following result.

Corollary 2.1.3 *Assume that*

$$a, b, A \in (0, \infty) \quad \text{and} \quad bA > a.$$

Then the positive equilibrium of Eq.(2.1.23) is globally asymptotically stable.

Computer observations suggest that when $a, b, A \in (0, \infty)$, the positive equilibrium $\bar{x}$ of Eq.(2.1.23) is globally asymptotically stable without any additional restriction on a, b and A. In this case, the only hypothesis of Theorem 2.1.1 which may not be satisfied by the mere assumption that $a, b, A \in (0, \infty)$ is that the function $uf(u, u)$ is increasing in u. Motivated by these observations we propose the following project.

Research Project 2.1.1 *Obtain a global asymptotic stability result for Eq.(2.1.1) without the assumption that $uf(u,u)$ is increasing in u and which includes as a special case the function f which is defined by (2.1.24) with $bA \leq a$.*

For more details on Eq.(2.1.23) see Sections 3.4 and 6.1.

Remark 2.1.1 It is interesting to note in which way the hypotheses of Theorem 2.1.1 fail to be true in each of the following three cases:

(a)

$$x_{n+1} = \frac{\alpha}{1 + \beta x_{n-1}}, \quad n = 0, 1, \ldots$$

where

$$\alpha \in (0, 1] \quad \text{and} \quad \beta \in (0, \infty).$$

Here as we will see in Theorem 3.4.1(a),

$$\lim_{n \to \infty} x_n = 0.$$

(b)

$$x_{n+1} = \frac{1 + x_n}{x_{n-1}}, \quad n = 0, 1, \ldots.$$

Here as we will see in Corollary 5.2.2, every positive solution is periodic with period five.

(c)

$$x_{n+1} = \frac{x_n}{x_{n-1}}, \quad n = 0, 1, \ldots.$$

Here every positive solution is periodic with period six.

2.2 Permanence of $x_{n+1} = x_n f(x_n, x_{n-k_1}, \ldots, x_{n-k_r})$

A difference equation

$$x_{n+1} = F(x_n, x_{n-1}, \ldots, x_{n-k}), \quad n = 0, 1, \ldots$$

is said to be **permanent** if there exist numbers C and D with $0 < C \leq D < \infty$ such that for any initial conditions $x_{-k}, \ldots, x_0 \in (0, \infty)$ there exists a positive integer N which depends on the initial conditions such that

$$C \leq x_n \leq D \quad \text{for} \quad n \geq N.$$

The importance of permanence for biological systems is thoroughly reviewed by Hutson and Schmitt [1].

Our aim in this section is to establish that under appropriate hypotheses the difference equation

$$x_{n+1} = x_n f(x_n, x_{n-k_1}, \ldots, x_{n-k_r}), \ n = 0, 1, \ldots \qquad (2.2.1)$$

is permanent.

Our result applies immediately to rational recursive sequences of the form

$$x_{n+1} = \frac{a + bx_n}{1 + \displaystyle\sum_{i=0}^{r} b_i x_{n-i}}, \ n = 0, 1, \ldots \qquad (2.2.2)$$

where $a, b, b_0, \ldots, b_r$ are nonnegative real numbers. See Section 3.1.

Throughout this section and the next, we will assume without further mention that $k_1, \ldots, k_r$ are positive integers and we will denote by k the maximum of $k_1, \ldots, k_r$. We will also assume that the function f satisfies the following hypotheses:

(H_1)

$$\left. \begin{array}{c} f \in C[(0, \infty) \times [0, \infty)^r, (0, \infty)] \\[2mm] \text{and} \quad g \in C[[0, \infty)^{r+1}, (0, \infty)] \end{array} \right\} \qquad (2.2.3)$$

where

$$g(u_0, u_1, \ldots, u_r) = u_0 f(u_0, u_1, \ldots, u_r)$$

$$\text{for} \quad u_0 \in (0, \infty) \quad \text{and} \quad u_1, \ldots, u_r \in [0, \infty), \quad \text{and}$$

$$g(0, u_1, \ldots, u_r) = \lim_{u_0 \to 0+} g(u_0, u_1, \ldots, u_r).$$

(H_2) $f(u_0, u_1, \ldots, u_r)$ *is nonincreasing in* $u_1, \ldots, u_r$*;*

(H_3) *The equation*

$$f(x, x, \ldots, x) = 1 \qquad (2.2.4)$$

has a unique positive solution $\bar{x}$*;*

(H_4) *Either the function* $f(u_0, u_1, \ldots, u_r)$ *does not depend on* u_0 *or for every* $x > 0$ *and* $u \geq 0$*,*

$$[f(x, u, \ldots, u) - f(\bar{x}, u, \ldots, u)](x - \bar{x}) \leq 0 \qquad (2.2.5)$$

with

$$[f(x, \bar{x}, \ldots, \bar{x}) - f(\bar{x}, \bar{x}, \ldots, \bar{x})](x - \bar{x}) < 0 \quad \text{for} \quad x \neq \bar{x}. \qquad (2.2.6)$$

If $a_{-k}, \ldots, a_{-1} \in [0, \infty)$ and $a_0 \in (0, \infty)$ are given, then Eq.(2.2.1) has a unique solution $\{x_n\}$ satisfying the initial conditions

$$x_j = a_j \quad \text{for} \quad j = -k, \ldots, 0.$$

Clearly

$$x_n > 0 \quad \text{for} \quad n \geq 0.$$

In the sequel, we will only consider solutions of Eq.(2.2.1) with initial values $a_{-k}, \ldots,$ $a_{-1} \in [0, \infty)$ and $a_0 \in (0, \infty)$.

Lemma 2.2.1 *Let $\{x_n\}$ be a solution of Eq.(2.2.1) such that for some $n_0 \geq 0$,*

$$\text{either} \qquad x_n \geq \bar{x} \quad \text{for} \quad n \geq n_0 \tag{2.2.7}$$

$$\text{or} \qquad x_n \leq \bar{x} \quad \text{for} \quad n \geq n_0. \tag{2.2.8}$$

Then for $n \geq n_0 + k$, the sequence $\{x_n\}$ is monotonic and

$$\lim_{n \to \infty} x_n = \bar{x}.$$

Proof. Assume that (2.2.7) holds. The case where (2.2.8) holds is similar and will be omitted. Then first by using (H$_2$) and then (H$_4$) we see that for $n \geq n_0 + k$,

$$x_{n+1} = x_n f(x_n, x_{n-k_1}, \ldots, x_{n-k_r}) \leq x_n f(x_n, \bar{x}, \ldots, \bar{x})$$

$$\leq x_n f(\bar{x}, \bar{x}, \ldots, \bar{x}) = x_n. \tag{2.2.9}$$

Hence $\{x_n\}$ is monotonic for $n \geq n_0 + k$. Let

$$l = \lim_{n \to \infty} x_n$$

and, for the sake of contradiction, assume that $l > \bar{x}$. Then by taking limits on both sides of Eq.(2.2.1), we obtain

$$f(l, l, \ldots, l) = 1$$

which contradicts the hypothesis that $\bar{x}$ is the only positive solution of Eq.(2.2.4). □

Corollary 2.2.1 *Every solution of Eq.(2.2.1), which is not strictly oscillatory about $\bar{x}$, tends to $\bar{x}$ as $n \to \infty$.*

The proof of the following lemma is similar to the part of the proof of Lemma 2.2.1 which led to (2.2.9) and therefore will be omitted.

Lemma 2.2.2 *Let $\{x_n\}$ be a solution of Eq.(2.2.1) which is strictly oscillatory about $\bar{x}$. Then the extreme point in any semicycle occurs in one of the first $k + 1$ terms of the semicycle.*

The main result in this section is the following:

Theorem 2.2.1 *Equation (2.2.1) is permanent.*

Proof. By the continuity of the function g, it follows that there exists a positive number A such that

$$xf(x, 0, \ldots, 0) \leq A \quad \text{for} \quad 0 \leq x \leq \bar{x}.$$

Set

$$D = \max\{2\bar{x}, A[f(\bar{x}, 0, \ldots, 0)]^k\}.$$

Again by the continuity of g, there exists a positive number B such that

$$xf(x, 0, \ldots, 0) \geq B \quad \text{for} \quad \bar{x} \leq x \leq D.$$

Set

$$C = \min\{\bar{x}/2, B[f(\bar{x}, D, \ldots, D)]^k\}.$$

Clearly

$$C < \bar{x} < D.$$

Let $\{x_n\}$ be a solution of Eq.(2.2.1). We now claim that for n sufficiently large,

$$C \leq x_n \leq D. \tag{2.2.10}$$

In view of Corollary 2.2.1, this is clearly true if $\{x_n\}$ is not strictly oscillatory about $\bar{x}$. So now assume that $\{x_n\}$ is strictly oscillatory about $\bar{x}$. Let $p \geq k$ and let

$$\{x_{p+1}, x_{p+2}, \ldots, x_q\}$$

be a positive semicycle followed by the negative semicycle

$$\{x_{q+1}, x_{q+2}, \ldots, x_s\}.$$

If x_M and x_m are the extreme values in these positive and negative semicycles, respectively, with the smallest possible indices M and m, then by Lemma 2.2.1,

$$M - p \leq k + 1 \quad \text{and} \quad m - q \leq k + 1. \tag{2.2.11}$$

Furthermore for any positive indices ν and N with $\nu < N$, by multiplying the equalities which result from Eq.(2.2.1) for $n = \nu, \ldots, N - 1$, we obtain

$$x_N = x_\nu \prod_{j=\nu}^{N-1} f(x_j, x_{j-k_1}, \ldots, x_{j-k_r}). \tag{2.2.12}$$

For $N = M$ and $\nu = p$, (2.2.12) yields

$$
\begin{aligned}
x_M &= x_p \prod_{j=p}^{M-1} f(x_j, x_{j-k_1}, \ldots, x_{j-k_r}) \\
&= x_p f(x_p, x_{p-k_1}, \ldots, x_{p-k_r}) \prod_{j=p+1}^{M-1} f(x_j, x_{j-k_1}, \ldots, x_{j-k_r}) \\
&\leq x_p f(x_p, 0, \ldots, 0)[f(\bar{x}, 0, \ldots, 0)]^{M-p-1} \\
&\leq x_p f(x_p, 0, \ldots, 0)[f(\bar{x}, 0, \ldots, 0)]^k < D. \qquad (2.2.13)
\end{aligned}
$$

Next for $N = m$ and $\nu = q$, (2.2.12) yields

$$
\begin{aligned}
x_m &= x_q \prod_{j=q}^{m-1} f(x_j, x_{j-k_1}, \ldots, x_{j-k_r}) \\
&\geq x_q f(x_q, D, \ldots, D) \prod_{j=q+1}^{m-1} f(x_j, D, \ldots, D) \\
&\geq x_q f(x_q, D, \ldots, D)[f(\bar{x}, D, \ldots, D)]^{m-q-1} \\
&\geq x_q f(x_q, D, \ldots, D)[f(\bar{x}, D, \ldots, D)]^k > C. \qquad (2.2.14)
\end{aligned}
$$

The proof is complete. □

Research Project 2.2.1 *Let $p \in \{1, 2, \ldots\}$. Obtain sufficient conditions under which the equation*

$$
x_{n+1} = x_n^p f(x_n, x_{n-k_1}, \ldots, x_{n-k_r}), \quad n = 0, 1, \ldots
$$

is permanent.

2.3 Global Attractivity of $x_{n+1} = x_n f(x_n, x_{n-k_1}, \ldots, x_{n-k_r})$

In this section we continue the study of Eq.(2.2.1) which was initiated in the previous section, and our goal is to establish a quite general global attractivity result for all positive solutions. As in the previous section, our results apply immediately to rational recursive sequences of the form of Eq.(2.2.2).

Throughout this section we assume without further mention that f satisfies the hypotheses $(H_1) - (H_4)$ stated in Section 2.2.

The next lemma establishes some useful properties of the function

$$F(x) = \begin{cases} \max_{x \leq y \leq \bar{x}} G(x,y) & \text{for} \quad 0 \leq x \leq \bar{x} \\[2ex] \min_{\bar{x} \leq y \leq x} G(x,y) & \text{for} \quad x > \bar{x} \end{cases} \qquad (2.3.1)$$

where

$$G(x,y) = yf(y,x,\ldots,x)f(\bar{x},\bar{x},\ldots,\bar{x},y)[f(\bar{x},x,\ldots,x)]^{k-1}. \qquad (2.3.2)$$

Lemma 2.3.1 $F \in C[(0,\infty),(0,\infty)]$ and F is nonincreasing in $[0,\infty)$.

Proof. From (2.3.2) and the standing hypotheses on f, it follows that the function $G(x,y)$ is continuous for $x,y \in [0,\infty)$ and nonincreasing in x for $x \geq 0$. Also clearly $F(\bar{x}) = G(\bar{x},\bar{x}) = \bar{x}$.

First we will prove that F is nonincreasing in $[0,\infty)$. Let $x_1 \in [0,x_2]$. If $x_2 \leq \bar{x}$, choose $y_2 \in [x_1,\bar{x}]$ such that

$$F(x_2) = \max_{x_2 \leq y \leq \bar{x}} G(x_2,y) = G(x_2,y_2).$$

Then

$$F(x_2) = G(x_2,y_2) \leq G(x_1,y_2) \leq \max_{x_1 \leq y \leq \bar{x}} G(x_1,y) = F(x_1).$$

Similarly if $x_1 \geq \bar{x}$, we can see that $F(x_1) \leq F(x_2)$. Finally, if $x_1 < \bar{x} < x_2$, then $F(x_2) \geq F(\bar{x}) \geq F(x_1)$ and the proof that F is nonincreasing is complete.

Next we prove that F is continuous. Otherwise there exists a point $x_0 \geq 0$ where F is discontinuous. We will assume that $x_0 \leq \bar{x}$. The case where $x_0 \geq \bar{x}$ is similar and will be omitted. Now if $x_0 = \bar{x}$, let $\{x_n\}$ be any sequence of points in $[0,\bar{x}]$ with limit $\bar{x}$. Also let $\{y_n\}$ be any sequence of points with

$$x_n \leq y_n \leq \bar{x} \quad \text{and} \quad F(x_n) = G(x_n,y_n).$$

Then $\lim_{n\to\infty} y_n = \bar{x}$, and since G is continuous,

$$\lim_{n\to\infty} F(x_n) = \lim_{n\to\infty} G(x_n,y_n) = G(\bar{x},\bar{x}) = \bar{x} = F(\bar{x})$$

which proves that F is continuous (from the left) at $\bar{x}$. So it remains to consider the case where the discontinuity of F is at a point $x_0 < \bar{x}$. Then there must exist $\epsilon_1 > 0$ such that for every $\delta_1 > 0$, there exists $x_1 > 0$ such that

$$|x_0 - x_1| < \delta_1 \quad \text{and} \quad |F(x_0) - F(x_1)| \geq \epsilon_1. \qquad (2.3.3)$$

On the other hand, as $G(x, z)$ is uniformly continuous on $[0, \bar{x}] \times [0, \bar{x}]$, for every $\epsilon_2 > 0$ there exists $\delta_2 = \delta_2(\epsilon_2) > 0$ such that

$$|x' - x''| < \delta_2 \quad \text{and} \quad |y' - y''| < \delta_2 \Rightarrow |G(x', y') - G(x'', y'')| < \epsilon_2.$$

Take $\epsilon_2 = \epsilon_1/2$, $\delta_1 = \delta_2$ and let $y_0 \in [x_0, \bar{x}]$ be such that $F(x_0) = G(x_0, y_0)$.

We will assume that $x_1 > x_0$. The case where $x_1 < x_0$ is similar and will be omitted. Now choose $y_1 \in [x_1, \bar{x}]$ such that $|y_1 - y_0| < \delta_1$. Then

$$G(x_0, y_0) - \epsilon_1/2 < G(x_1, y_1) < G(x_0, y_0) + \epsilon_1/2.$$

But

$$F(x_1) \geq G(x_1, y_1) > G(x_0, y_0) - \epsilon_1/2 = F(x_0) - \epsilon_1/2$$

and so

$$0 \leq F(x_0) - F(x_1) \leq \epsilon_1/2$$

which contradicts (2.3.3) and completes the proof. $\qquad\square$

In addition to the standing hypotheses on f, which were assumed in Section 2.2, we will now also assume that

(H_5) $f(u_0, u_1, \ldots, u_r)$ is nonincreasing in u_0.

However a simple analysis of the proof in this section will reveal that instead all we need to assume is the following weaker condition:

(H_5')

$$\left. \begin{array}{l} f(x, \bar{x}, \ldots, \bar{x}, y) \leq f(\bar{x}, \bar{x}, \ldots, \bar{x}, y) \quad \text{for} \quad x \geq \bar{x} \text{ and } 0 \leq y \leq \bar{x} \\[2mm] f(x, \bar{x}, \ldots, \bar{x}, y) \geq f(\bar{x}, \bar{x}, \ldots, \bar{x}, y) \quad \text{for} \quad 0 < x \leq \bar{x} \text{ and } y \geq \bar{x}. \end{array} \right\}$$

Theorem 2.3.1 *Assume that the function F which was defined by (2.3.1) has no periodic points of prime period 2. Then $\bar{x}$ is a global attractor of all positive solutions of Eq.(2.2.1).*

Proof. Let $\{x_n\}$ be a positive solution of Eq.(2.2.1). By Lemma 2.2.1, if $\{x_n\}$ is not strictly oscillatory then,

$$\lim_{n \to \infty} x_n = \bar{x}. \tag{2.3.4}$$

So it remains to establish (2.3.4) when $\{x_n\}$ is strictly oscillatory. To this end, let

$$\{x_{p_i+1}, x_{p_i+2}, \ldots, x_{q_i}\}$$

be the ith positive semicycle of $\{x_n\}$ followed by the ith negative semicycle

$$\{x_{q_i+1}, x_{q_i+2}, \ldots, x_{p_{i+1}}\}.$$

Let x_{M_i} and x_{m_i} be the extreme values in these two semicycles, respectively, with the smallest possible indices M_i and m_i. Then as in the proof of Theorem 2.2.1,

$$M_i - p_i \le k + 1, \ m_i - q_i \le k + 1, \tag{2.3.5}$$

$$x_{M_i} = x_{p_i} \prod_{j=p_i}^{M_i-1} f(x_j, x_{j-k_1}, \ldots, x_{j-k_r}) \tag{2.3.6}$$

and

$$x_{m_i} = x_{q_i} \prod_{j=q_i}^{m_i-1} f(x_j, x_{j-k_1}, \ldots, x_{j-k_r}). \tag{2.3.7}$$

Let

$$\left.\begin{array}{rcl}
\lambda &=& \displaystyle\liminf_{n\to\infty} x_n = \liminf_{i\to\infty} x_{m_i} \\[2mm]
\mu &=& \displaystyle\limsup_{n\to\infty} x_n = \limsup_{i\to\infty} x_{M_i}
\end{array}\right\} \tag{2.3.8}$$

which in view of (2.2.10) exist and are such that

$$0 < C \le \lambda \le \bar{x} \le \mu < D.$$

To complete the proof it suffices to show that

$$\lambda = \mu = \bar{x}. \tag{2.3.9}$$

From (2.3.8) it follows that if $\eta \in (0, \infty)$ and $\epsilon \in (0, \lambda)$ are given, then there exists $n_0 \in \mathbf{N}$ such that

$$\lambda - \epsilon \le x_n \le \mu \quad \text{for} \quad n \ge n_0 - k. \tag{2.3.10}$$

We now claim that

$$x_{M_i} \le G(\lambda - \epsilon, x_{p_i}) \tag{2.3.11}$$

where G is given by (2.3.2). Indeed, if

$$M_i - p_i < k + 1 \tag{2.3.12}$$

then

$$\begin{aligned}
x_{M_i} &= x_{p_i} \prod_{j=p_i}^{M_i-1} f(x_j, x_{j-k_1}, \ldots, x_{j-k_r}) \\[2mm]
&\le x_{p_i} f(x_{p_i}, \lambda - \epsilon, \ldots, \lambda - \epsilon) \prod_{j=p_i+1}^{M_i-1} f(x_j, \lambda - \epsilon, \ldots, \lambda - \epsilon).
\end{aligned}$$

Furthermore, by using (2.2.5) and (2.3.12) we see that

$$x_{M_i} \leq x_{p_i} f(x_{p_i}, \lambda - \epsilon, \ldots, \lambda - \epsilon)[f(\bar{x}, \lambda - \epsilon, \ldots, \lambda - \epsilon)]^{M_i - p_i}$$

$$\leq x_{p_i} f(x_{p_i}, \lambda - \epsilon, \ldots, \lambda - \epsilon)[f(\bar{x}, \lambda - \epsilon, \ldots, \lambda - \epsilon)]^{k-1}.$$

But $f(\bar{x}, \bar{x}, \ldots, x_{p_i}) \geq f(\bar{x}, \bar{x}, \ldots, \bar{x}) = 1$ and so

$$x_{M_i} \leq x_{p_i} f(x_{p_i}, \lambda - \epsilon, \ldots, \lambda - \epsilon)[f(\bar{x}, \lambda - \epsilon, \ldots, \lambda - \epsilon)]^{k-1}$$

$$\times \ f(\bar{x}, \bar{x}, \ldots, x_{p_i}) = G(\lambda - \epsilon, x_{p_i})$$

and (2.3.11) is established when (2.3.12) holds. On the other hand if (2.3.12) does not hold, then $M_i - p_i = k + 1$ and in this case,

$$x_{M_i} = x_{p_i} f(x_{p_i}, x_{p_i - k_1}, \ldots, x_{p_i - k_r})[\prod_{j = p_i + 1}^{M_i - 2} f(x_j, x_{j - k_1}, \ldots, x_{j - k_r})]$$

$$\times \ f(x_{M_i - 1}, x_{M_i - 1 - k_1}, \ldots, x_{M_i - 1 - k_r})$$

$$\leq x_{p_i} f(x_{p_i}, \lambda - \epsilon, \ldots, \lambda - \epsilon)[f(\bar{x}, \lambda - \epsilon, \ldots, \lambda - \epsilon)]^{k-1}$$

$$\times \ f(\bar{x}, \bar{x}, \ldots, x_{p_i}) = G(\lambda - \epsilon, x_{p_i})$$

which completes the proof that (2.3.11) holds. Since $\lambda - \epsilon < x_{p_i} \leq \bar{x}$, it follows from (2.3.11) that

$$x_{M_i} \leq G(\lambda - \epsilon, x_{p_i}) \leq \max_{\lambda - \epsilon \leq y \leq \bar{x}} G(\lambda - \epsilon, y) = F(\lambda - \epsilon).$$

Therefore, as ϵ is arbitrary, $x_{M_i} \leq F(\lambda)$ and so from (2.3.8)

$$\mu \leq F(\lambda). \tag{2.3.13}$$

In a similar way we can show that

$$\lambda \geq F(\mu). \tag{2.3.14}$$

By applying Lemma 1.4.3, it finally follows that (2.3.9) is true and the proof is complete. $\square$

The main difficulty in applying Theorem 2.3.1 is showing that the function F has no periodic orbits of prime period 2. The next result establishes the global attractivity of the positive equilibrium $\bar{x}$ of Eq.(2.2.1) under conditions which although stronger than those of Theorem 2.3.1 are, however, easier to verify.

Theorem 2.3.2 *Suppose that either condition (a) or (b) given below is satisfied:*

(a) Assume either that the function $f(u_0, u_1, \ldots, u_r)$ does not depend on u_0, or that for every $x, y \in (0, \infty)$ with $x \neq \bar{x}$,

$$[xf(x, y, \ldots, y) - \bar{x}f(\bar{x}, y, \ldots, y)](x - \bar{x}) > 0. \tag{2.3.15}$$

Furthermore assume that the function F_1 which is given by

$$F_1(x) = \bar{x}f(\bar{x}, x, \ldots, x)^{k+1} \tag{2.3.16}$$

has no periodic orbits of prime period 2.

(b) Assume that for every $x, y \in (0, \infty)$ with $x \neq \bar{x}$,

$$[xf(x, y, \ldots, y)f(\bar{x}, \ldots, \bar{x}, x) - \bar{x}f(\bar{x}, y, \ldots, y)](x - \bar{x}) > 0 \tag{2.3.17}$$

and that the function F_2 given by

$$F_2(x) = \bar{x}f(\bar{x}, x, \ldots, x)^{k} \tag{2.3.18}$$

has no periodic orbits of prime period 2.

Then $\bar{x}$ is a global attractor of all positive solutions of Eq.(2.2.1).

Proof. Assume that (a) holds. Let $0 < x < \bar{x}$ and $x \leq y \leq \bar{x}$. Then from (2.3.15) it follows that

$$xf(x, y, \ldots, y) \leq \bar{x}f(\bar{x}, y, \ldots, y).$$

Furthermore, by using the monotonic character of f we see that

$$f(\bar{x}, \ldots, \bar{x}, y) \leq f(\bar{x}, x, \ldots, x).$$

From the above and (2.3.2) we obtain

$$G(x, y) \leq \bar{x}f(\bar{x}, x, \ldots, x)^{k+1},$$

$$F(x) \leq F_1(x) \quad \text{for} \quad 0 < x < \bar{x},$$

and

$$F(x) \geq F_1(x) \quad \text{for} \quad x > \bar{x}.$$

Then, from Lemma 1.4.4 it follows that the only positive solution of equation

$$F^2(x) = x$$

is $x = \bar{x}$. Hence by Theorem 2.3.1, $\bar{x}$ is a global attractor of all positive solutions of Eq.(2.2.1).

Assume that (b) holds. Then from (2.3.17), for $0 < x < \bar{x}$ and $x \le y \le \bar{x}$, it follows that

$$xf(x,y,\ldots,y)f(\bar{x},\ldots,\bar{x},x) \le \bar{x}f(\bar{x},y,\ldots,y).$$

Also from (2.3.2) we find

$$G(x,y) \le \bar{x}f(\bar{x},x,\ldots,x)^k$$

which implies

$$F(x) \le F_2(x) \quad \text{for} \quad 0 < x < \bar{x}.$$

Furthermore we have

$$F(x) \ge F_2(x) \quad \text{for} \quad x > \bar{x}$$

from which the conclusion follows in a way similar to the above. □

The following corollary was originally established in Kocic and Ladas [3].

Corollary 2.3.1 *Consider the difference equation*

$$x_{n+1} = x_n h(x_{n-k}), \ n = 0,1,\ldots \qquad (2.3.19)$$

where k is a positive integer and h satisfies the following properties:

(a) $h \in C[[0,\infty),(0,\infty)]$ and $h(u)$ is nonincreasing in u;

(b) The equation $h(x) = 1$ has a unique positive solution;

(c) If $\bar{x}$ denotes the unique positive solution of $h(x) = 1$, then

$$[xh(x) - \bar{x}](x - \bar{x}) > 0 \quad \text{for} \quad x \ne \bar{x};$$

(d) The only solution of the equation

$$\bar{x}[h(\bar{x}(h(x))^k)]^k = x$$

in the interval $0 \le x \le \bar{x}(h(0))^k$, is $x = \bar{x}$.

Then $\bar{x}$ is a global attractor of all positive solutions of Eq.(2.3.19).

Remark 2.3.1 One can see that when $k = 1$, condition (d) in Corollary 2.3.1 is a consequence of conditions (a)-(c). Also conditions (c) and (d) can be replaced by the following condition:

(e) The only solution of the equation

$$\bar{x}[h(\bar{x}(h(x))^{k+1})]^{k+1} = x$$

in the interval $0 \le x \le \bar{x}(h(0))^{k+1}$ is $x = \bar{x}$.

Research Project 2.3.1 *Obtain global attractivity results for equations of the form*

$$x_{n+1} = x_n^p f(x_n, x_{n-k_1}, \ldots, x_{n-k_r}), \; n = 0, 1, \ldots$$

where p is a positive integer greater than one. For motivation consider the rational recursive sequence

$$x_{n+1} = \frac{a + bx_n^2}{1 + x_{n-1}^2}, \; n = 0, 1, \ldots$$

where $a, b \in [0, \infty)$ for which computer observations indicate that for the certain values of a and b the positive equilibrium is a global attractor of all positive solutions.

2.4 Global Attractivity of $x_{n+1} = \alpha x_n + F(x_{n-k})$

The main result in this section is the following global attractivity result of Karakostas, Philos and Sficas [1] for the difference equation

$$x_{n+1} = \alpha x_n + F(x_{n-k}), \; n = 0, 1, \ldots \tag{2.4.1}$$

where

$$\alpha \in [0, 1), \; k \in \{1, 2, \ldots\} \quad \text{and} \quad F \in C[[0, \infty), (0, \infty)]. \tag{2.4.2}$$

For applications of this result see Section 4.6.

Theorem 2.4.1 *Assume that (2.4.2) holds, $F(u)$ is decreasing in u, and that the system*

$$U = \frac{F(L)}{1 - \alpha} \quad \text{and} \quad L = \frac{F(U)}{1 - \alpha} \tag{2.4.3}$$

has exactly one solution $\{L, U\}$ in the positive quadrant $(0, \infty) \times (0, \infty)$. Then Eq.(2.4.1) has a unique positive equilibrium $\bar{x}$. Furthermore,

$$U = L = \bar{x}$$

and every solution $\{x_n\}$ of Eq.(2.4.1) with positive initial conditions

$$x_n > 0 \quad \text{for} \quad n = -k, \ldots, 0$$

is attracted to $\bar{x}$; that is,

$$\lim_{n \to \infty} x_n = \bar{x}. \tag{2.4.4}$$

Proof. Clearly $\bar{x}$ is a positive equilibrium of Eq.(2.4.1) if and only if $\bar{x}$ is a positive solution of the equation

$$U = \frac{F(U)}{1-\alpha} \tag{2.4.5}$$

which in turn is true if and only if $\{\bar{x}, \bar{x}\}$ is a positive solution of (2.4.3) in the positive quadrant. Hence Eq.(2.4.1) has at most one positive equilibrium. On the other hand, Eq.(2.4.5) has at least one positive solution because if we set

$$G(u) = u - \frac{F(u)}{1-\alpha},$$

then

$$G(0) < 0 \quad \text{and} \quad G(\infty) = \infty.$$

Therefore Eq.(2.4.1) has a unique positive equilibrium $\bar{x}$. Also $U = L = \bar{x}$ because $\{\bar{x}, \bar{x}\}$ is a solution of (2.4.3), and (2.4.3) has exactly one solution. It remains to establish (2.4.4). To this end, observe that

$$x_{n+1} \le \alpha x_n + F(0), \quad n = 0, 1, \ldots$$

and so by Lemmas 1.4.1 and 1.4.2,

$$\limsup_{n\to\infty} x_n \le \frac{F(0)}{1-\alpha}.$$

Set

$$U_1 = \frac{F(0)}{1-\alpha}, \quad L_1 = \frac{F(U_1)}{1-\alpha}$$

and for $m = 1, 2, \ldots$, set

$$U_{m+1} = \frac{F(L_m)}{1-\alpha}, \quad \text{and} \quad L_{m+1} = \frac{F(U_{m+1})}{1-\alpha}.$$

Now we can see by induction that $\{U_m\}$ is a decreasing sequence, $\{L_m\}$ is an increasing sequence, and that for $m = 1, 2, \ldots$

$$L_m \le \liminf_{n\to\infty} x_n \le \limsup_{n\to\infty} x_n \le U_m.$$

Set

$$L = \lim_{m\to\infty} L_m \quad \text{and} \quad U = \lim_{m\to\infty} U_m.$$

Then $\{L, U\}$ is a solution of Eq.(2.4.3) in the positive quadrant. Therefore

$$U = L = \bar{x}$$

and the proof is complete. $\square$

By using Lemma 1.6.3 we obtain the following interesting corollary of Theorem 2.4.1.

Corollary 2.4.1 *Assume that* $\alpha \in [0,1)$, $k \in \{1,2,\dots\}$, *and* $F \in C[[0,\infty),(0,\infty)]$ *is decreasing. Suppose that F has a unique fixed point $\bar{x} \in (0,\infty)$ and that $\bar{x}$ is a global attractor of all positive solutions of the first order difference equation*

$$y_{n+1} = \frac{F(y_n)}{1-\alpha}, \ n = 0,1,\dots \tag{2.4.6}$$

where $y_0 \in (0,\infty)$. Then $\bar{x}$ is a global attractor of all positive solutions of Eq.(2.4.1).

Research Project 2.4.1 *Extend Theorem 2.4.1 to equations of the form*

$$X_{n+1} = AX_n + F(X_{n-k}), \ n = 0,1,\dots$$

where A is an $m \times m$ matrix and $F \in C[(0,\infty)^m,(0,\infty)^m]$.

2.5   Global Stability of
$$x_{n+1} = \sum_{i=0}^{k} a_i x_{n-i} + (1 - A)F\left(\sum_{i=0}^{m} b_i x_{n-i}\right)$$

In this section we study the global stability of the delay difference equation

$$x_{n+1} = \sum_{i=0}^{k} a_i x_{n-i} + (1-A)F\left(\sum_{i=0}^{m} b_i x_{n-i}\right), \ n = 0,1,\dots \tag{2.5.1}$$

where

$$\left.\begin{array}{l} a_0,\dots,a_k,b_0,\dots,b_m \in [0,\infty), \quad r = \max\{k,m\}, \\[2mm] A = \displaystyle\sum_{i=0}^{k} a_i < 1, \quad \sum_{i=0}^{m} b_i = 1 \end{array}\right\} \tag{2.5.2}$$

and

$$\left.\begin{array}{l} F \in C[[0,\infty),[0,\infty)]], \quad F \text{ has a unique positive fixed point } \bar{x}, \\[2mm] \text{and} \quad (x-\bar{x})(F(x)-x) < 0 \quad \text{for} \quad 0 < x \neq \bar{x}. \end{array}\right\} \tag{2.5.3}$$

Equations of this type have been employed by Clark [1] in the study of whale populations. See Section 4.7.

Theorem 2.5.1 *Assume that (2.5.2) and (2.5.3) hold. Let $\{x_n\}$ be a solution of Eq.(2.5.1) with positive initial conditions $x_{-r},\dots,x_0$. Then the following statements are true:*

(a) Let

$$P = \max\{x_{-r}, \ldots, x_0, \max_{0 \le x \le \bar{x}}\{F(x)\}\},$$

$$Q = \min\{x_{-r}, \ldots, x_0, \min_{\bar{x} \le x \le P}\{F(x)\}\}. \Bigg\} \tag{2.5.4}$$

Then

$$0 < Q \le x_n \le P \quad \text{for} \quad n = -r, -r+1, \ldots. \tag{2.5.5}$$

(b) Assume that F is also a monotonic function and suppose that $\bar{x}$ is a global attractor of all positive solutions of the first-order difference equation

$$z_{n+1} = F(z_n), \ n = 0, 1, \ldots. \tag{2.5.6}$$

Then $\bar{x}$ is a global attractor of all positive solutions of Eq.(2.5.1).

Proof. (a) The proof is by induction. Clearly (2.5.5) holds for $n = -r, \ldots, 0$. Suppose that (2.5.5) holds for all integers less than or equal to n. Then

$$x_{n+1} = \sum_{i=0}^{k} a_i x_{n-i} + (1-A)F\left(\sum_{i=0}^{m} b_i x_{n-i}\right)$$

$$\le \left(\sum_{i=0}^{k} a_i\right)P + (1-A)C \le AP + (1-A)P = P$$

and the second inequality in (2.5.5) holds. First, consider the case where

$$\bar{x} \le \sum_{i=0}^{m} b_i x_{n-i} \le P.$$

Then we have

$$F\left(\sum_{i=0}^{m} b_i x_{n-i}\right) \ge \min_{\bar{x} \le x \le P}\{F(x)\} \ge Q$$

and

$$x_{n+1} = \sum_{i=0}^{k} a_i x_{n-i} + (1-A)F\left(\sum_{i=0}^{m} b_i x_{n-i}\right)$$

$$\ge AQ + (1-A)Q = Q.$$

Next consider the case where

$$0 < \sum_{i=0}^{m} b_i x_{n-i} < \bar{x}.$$

Then from (2.5.3) it follows that

$$F\left(\sum_{i=0}^{m} b_i x_{n-i}\right) > \sum_{i=0}^{m} b_i x_{n-i} \geq Q$$

and, again, the first inequality in (2.5.5) holds.
(b) Let $\{x_n\}$ be a solution of Eq.(2.5.1) and set

$$\lambda = \liminf_{n \to \infty} x_n \quad \text{and} \quad \mu = \limsup_{n \to \infty} x_n \qquad (2.5.7)$$

which exist because the solution is bounded. From (a) it follows that

$$0 < \lambda \leq x_n \leq \mu < \infty. \qquad (2.5.8)$$

To complete the proof it suffices to show that

$$\lambda = \mu = \bar{x}. \qquad (2.5.9)$$

From (2.5.7) it follows that if $\eta \in (0, \infty)$ and $\epsilon \in (0, \lambda)$ are given, then there exists $n_0 \in \mathbb{N}$ such that

$$\lambda - \epsilon \leq x_n \leq \mu + \eta \quad \text{for} \quad n \geq n_0 - r. \qquad (2.5.10)$$

Furthermore for $n \geq n_0$, we have

$$A(\lambda - \epsilon) \leq \sum_{i=0}^{k} a_i x_{n-i} \leq A(\mu + \eta)$$

and

$$\lambda - \epsilon \leq \sum_{i=0}^{m} b_i x_{n-i} \leq \mu + \eta.$$

First consider the case where F is a nonincreasing function. Then from the above it follows that

$$\left.\begin{array}{rl} x_{n+1} \leq & A(\mu + \eta) + (1 - A)F(\lambda + \epsilon), \\ \\ x_{n+1} \geq & A(\lambda + \epsilon) + (1 - A)F(\mu + \eta). \end{array}\right\} \qquad (2.5.11)$$

Therefore, since ϵ and μ are arbitrary, from (2.5.7) and (2.5.11) we obtain

$$\mu \leq F(\lambda) \quad \text{and} \quad \lambda \geq F(\mu). \qquad (2.5.12)$$

Since $\bar{x}$ is a global attractor of all positive solutions of Eq.(2.5.6), it follows that $\bar{x}$ is the only fixed point of $F^2(x)$ in $(0, \infty)$. Then by applying Lemma 1.4.3, we conclude that (2.5.9) holds.

Next consider the case where F is nondecreasing. By an argument similar to the above we find

$$\mu \le F(\mu) \quad \text{and} \quad \lambda \ge F(\lambda).$$

Then from (2.5.3) and (2.5.8), we again conclude that (2.5.9) holds. The proof is complete. □

The following Theorem gives sufficient conditions for the global asymptotic stability of Eq.(2.5.1).

Theorem 2.5.2 *Assume that (2.5.2) and (2.5.3) hold and suppose that there exists a convex function v which is a Liapunov function of the first-order difference equation (2.5.6) on $(0, \infty)$, such that $v(F(x)) < v(x)$ for $x \ne \bar{x}$. Then the positive equilibrium $\bar{x}$ of Eq.(2.5.1) is globally asymptotically stable.*

Proof. By introducing the substitution

$$y_n^i = x_{n-i} \quad \text{for} \quad i = 1, \ldots, r+1 \quad \text{and} \quad n = 0, 1, \ldots$$

we see that Eq.(2.5.1) is equivalent to the system,

$$
\begin{aligned}
y_{n+1}^1 &= y_n^2 \\
&\vdots \\
y_{n+1}^r &= y_n^{r+1} \\
y_{n+1}^{r+1} &= \sum_{i=0}^{k} a_i y_n^{r+1-i} + (1-A) F\left(\sum_{i=0}^{m} b_i y_n^{r+1-i} \right).
\end{aligned}
\tag{2.5.13}
$$

Let $g_i \in C[[0, \infty)^{r+1}, [0, \infty)]$ for $i = 1, \ldots, r+1$ be functions which are defined as follows: For $U = [u_1, \cdots, u_{r+1}]^T \in [0, \infty)^{r+1}$, set

$$g_i(U) = u_{i+1} \quad \text{for} \quad i = 1, \ldots, r$$

and

$$g_{r+1}(U) = \sum_{i=0}^{k} a_i u_{r+1-i} + (1-A) F\left(\sum_{i=0}^{m} b_i u_{r+1-i} \right).$$

Furthermore, let G be the vector-valued function defined by

$$G(U) = [g_1(U), \cdots, g_{r+1}(U)]^T].$$

Then the system (2.5.13) can be written in the form

$$Y_{n+1} = G(Y_n), \quad n = 0, 1, \ldots \tag{2.5.14}$$

where $Y_n = [y_n^1, \cdots, y_n^{r+1}]^T$ for $n = 0, 1, \ldots$. Since v is a convex function and a Liapunov function of the first-order equation Eq.(2.5.6), we have

$$v(g_i(U)) = v(u_{i+1}) \quad \text{for} \quad i = 1, \ldots, r$$

and

$$
\begin{aligned}
v(g_{r+1}(U)) &= v\Big(\sum_{i=0}^{k} a_i u_{r+1-i} + (1-A)F\Big(\sum_{i=0}^{m} b_i u_{r+1-i}\Big)\Big) \\
&\leq \sum_{i=0}^{k} a_i v(u_{r+1-i}) + (1-A)v\Big(F\Big(\sum_{i=0}^{m} b_i u_{r+1-i}\Big)\Big) \\
&\leq \sum_{i=0}^{k} a_i v(u_{r+1-i}) + (1-A)v\Big(\sum_{i=0}^{m} b_i u_{r+1-i}\Big) \\
&\leq \sum_{i=0}^{k} a_i v(u_{r+1-i}) + (1-A)\sum_{i=0}^{m} b_i v(u_{r+1-i}) \\
&\leq \max\{v(u_1), \ldots, v(u_{r+1})\}.
\end{aligned}
$$

Now define the function V by setting

$$V(U) = \max\{v(u_1), \ldots, v(u_{r+1})\}. \tag{2.5.15}$$

As v is a continuous function, V is also continuous. Clearly,

$$
\begin{aligned}
V(G(U)) &= \max\{v(g_1(U)), \ldots, v(g_{r+1}(U))\} \\
&\leq \max\{v(u_2), \ldots, v(u_{r+1}), \max\{v(u_1), \ldots, v(u_{r+1})\}\} \\
&= \max\{v(u_1), \ldots, v(u_{r+1})\} = V(U)
\end{aligned}
$$

and so V is a Liapunov function for the system (2.5.14). The only point possible in the boundary of the set of positive vectors in $\mathbf{R}^{r+1}$ where $V(G(X)) = V(X)$ is the point $\mathbf{0}$.

If $F(0) \neq 0$, then there are no such points and from Theorem 1.3.1(a) it follows that $\bar{X} = [\bar{x}, \ldots, \bar{x}]^T$ is globally asymptotically stable. Hence $\bar{x}$ is also a globally asymptotically stable positive equilibrium of Eq.(2.5.1).

Next assume $F(0) = 0$. Since the solutions of Eq.(2.5.1) are bounded away from zero it follows that no solution of (2.5.14) with positive initial conditions can approach $\mathbf{0}$. Hence from Theorem 1.3.1 $\bar{X}$ is globally asymptotically stable, and so $\bar{x}$ is a globally asymptotically stable positive equilibrium of Eq.(2.5.1). The proof is complete.   □

2.6   Notes

For some interesting results on the dynamics of second order difference equations see Marotto [1, 2].

The results in Section 2.1 are improved versions of the results in Jaroma, Kocic and Ladas [1]. See also Kocic and Ladas [4]. The results in Sections 2.2 and 2.3 are from Kocic and Ladas [5].

Theorem 2.4.1 is extracted from Karakostas, Philos and Sficas [1]. For some related work see Ivanov [1] where he investigated the nonlinear difference equation

$$\mu \Delta x_n = -x_{n+1} + f(x_{n-k}), \ n = 0, 1, \ldots \qquad (2.6.1)$$

where $\mu \in [0, \infty)$, $\Delta x_n = x_{n+1} - x_n$, $f \in C[I, I]$ for some closed interval I, and $k \in \{1, 2, \ldots\}$. He established the following result.

Theorem 2.6.1 *Assume that the difference equation*

$$x_{n+1} = f(x_n), \ n = 0, 1, \ldots \qquad (2.6.2)$$

has a unique equilibrium $\bar{x}$ and that $\bar{x}$ is a global attractor of all solutions of Eq.(2.6.2) with x_0 in the interior of I. Then $\bar{x}$ is a global attractor of all solutions of Eq.(2.6.1) with initial conditions $x_{-k}, \ldots, x_0$ in the interior of I.

The results in Section 2.5 are extracted from Fisher [1] and Fisher and Goh [1]. The global attractivity of the positive equilibrium of the difference equation

$$x_{n+1} = a + \sum_{k=0}^{m} \frac{b_k}{x_{n-k}}, \ n = 0, 1, \ldots$$

where a and $b_0, \ldots, b_m$ are nonnegative numbers was investigated by Philos, Purnaras and Sficas [1].

Finally we present the following result of Hautus and Bolis [1] which although we could not fit in any of the sections of this chapter it is interesting in its own right.

Theorem 2.6.2 *Consider the difference equation*

$$x_{n+1} = g(x_n, \ldots, x_{n-k}), \ n = 0, 1, \ldots \qquad (2.6.3)$$

where $g \in C[(0, \infty)^{k+1}, (0, \infty)]$ is increasing in each of its arguments and where the initial conditions $x_{-k}, \ldots, x_0$ are positive. Assume that Eq.(2.6.3) has a unique positive equilibrium $\bar{x}$ and suppose that the function h defined by

$$h(x) = g(x, \ldots, x), \ x \in (0, \infty)$$

satisfies

$$(h(x) - x)(x - \bar{x}) < 0 \quad \text{for} \quad x \neq \bar{x}.$$

Then $\bar{x}$ is a global attractor of all positive solutions of Eq.(2.6.3).

Proof. Let

$$m = \min\{x_{-k}, \ldots, x_0, \bar{x}\} \quad \text{and} \quad M = \max\{x_{-k}, \ldots, x_0, \bar{x}\}.$$

Clearly, $0 < m \le \bar{x} \le M < \infty$ and for $n = -k, \ldots 0,$

$$m \le x_n \le M. \tag{2.6.4}$$

Suppose that (2.6.4) holds for all $n \le N$. Then

$$x_{N+1} = g(x_N, \ldots, x_{N-k}) \le g(M, \ldots, M) = h(M) \le M.$$

Similarly we have $x_{N+1} \ge m$ and by induction (2.6.4) holds for all $n \ge -k$. Let

$$\lambda = \liminf_{n \to \infty} x_n \quad \text{and} \quad \mu = \limsup_{n \to \infty} x_n.$$

Then

$$\lambda = \liminf_{n \to \infty} g(x_n, \ldots, x_{n-k}) \ge g(\lambda, \ldots, \lambda) = h(\lambda)$$

and so $\lambda \ge \bar{x}$. In a similar way we find that $\mu \le \bar{x}$ and the proof is complete. □

Chapter 3

Rational Recursive Sequences

Our goal in this chapter is to gain some understanding of the dynamics of rational recursive sequences of the form

$$x_{n+1} = \frac{a + \sum_{i=0}^{k} a_i x_{n-i}}{b + \sum_{i=0}^{k} b_i x_{n-i}}, \quad n = 0, 1, \ldots \tag{3.0.1}$$

where k is a positive integer and the coefficients a, b, $a_0, \ldots, a_k$, $b_0, \ldots, b_k$ are nonnegative real numbers.

When $k = 0$, Eq.(3.0.1) reduces to the first order difference equation

$$x_{n+1} = \frac{a + a_0 x_n}{b + b_0 x_n}, \quad n = 0, 1, \ldots \tag{3.0.2}$$

called the **Riccati difference equation** which has been thoroughly investigated by Brand [1]. See also Saaty [1] for an application of Eq.(3.0.2) to optics. Eq.(3.0.2) and the more general Riccati difference equation with variable coefficients,

$$x_{n+1} = \frac{a_n x_n + b_n}{c_n x_n + d_n}, \quad n = 0, 1, \ldots \tag{3.0.3}$$

has the luxurious property, not shared by the general Eq.(3.0.1), that it can be transformed to a linear second order difference equation. For the sake of completeness and with an eye towards generalizations to equations with variable coefficients, we present in Appendix A a wealth of information about the solutions of Eq.(3.0.3).

3.1   The Rational Recursive Sequence
$$x_{n+1} = (a + bx_n)/\left(1 + \sum_{i=0}^{k} a_i x_{n-i}\right)$$

In this section we investigate the global attractivity of all positive solutions of the
difference equation

$$x_{n+1} = \frac{a + bx_n}{1 + \sum_{i=0}^{k} b_i x_{n-i}}, \quad n = 0, 1, \ldots \tag{3.1.1}$$

where

$$a, b_0, \ldots, b_{k-1} \in [0, \infty) \quad \text{and} \quad b, b_k \in (0, \infty). \tag{3.1.2}$$

Eq.(3.1.1) has a unique positive equilibrium $\bar{x}$ provided that

$$\left.\begin{array}{ll} \text{either} & a > 0 \\[2mm] \text{or} & a = 0 \text{ and } b > 1. \end{array}\right\} \tag{3.1.3}$$

Furthermore, $\bar{x}$ is given by

$$\bar{x} = \frac{b - 1 + \sqrt{(b-1)^2 + 4aB}}{2B} \tag{3.1.4}$$

where

$$B = \sum_{i=0}^{k} b_i > 0. \tag{3.1.5}$$

When $a = 0$ and $0 < b \le 1$, zero is an equilibrium of Eq.(3.1.1) and there is no
positive equilibrium.

The following result describes the asymptotic stability of all positive solutions of
Eq.(3.1.1).

Theorem 3.1.1 *Assume that (3.1.2) holds. Then the following statements are true.*

(a) *Assume that $a = 0$ and $0 < b \le 1$. Then every positive solution of Eq.(3.1.1)
decreases to zero.*

(b) *Assume that (3.1.3) holds. Then Eq.(3.1.1) is permanent.*

(c) *Assume that (3.1.3) holds and that one of the following three conditions is satisfied:*

(i) $B\bar{x}k \leq 1;$

(ii) $(B - b_0)\bar{x}k \leq 1 + b_0\bar{x}$ *and* $b - ab_0 \geq 0;$

(iii) $B\bar{x}(k - 1) \leq 1$ *and* $b[1 + \bar{x}(B - b_k)] - b_k a \geq 0.$

Then $\bar{x}$ is a global attractor of all positive solutions of Eq.(3.1.1).

Proof. (a) Clearly

$$x_{n+1} \leq bx_n \leq x_n$$

and so $x = \lim_{n\to\infty} x_n$ exists and is nonnegative. Then

$$x = \frac{bx}{1 + Bx}$$

from which it follows that $x = 0$.

(b) This result follows by observing that the function

$$f(u_0, u_1, \ldots, u_k) = \frac{b + \dfrac{a}{u_0}}{1 + \displaystyle\sum_{i=0}^{k} b_i u_i}$$

satisfies all the hypotheses of Theorem 2.2.1.

(c) We will apply Theorems 2.3.1 and 2.3.2. Here the function G is given by

$$G(x, y) = \frac{(a + by)\left(b + \dfrac{a}{\bar{x}}\right)^k}{(1 + b_0 y + Cx)(1 + D\bar{x} + b_r y)(1 + b_0\bar{x} + Cx)^{k-1}} \tag{3.1.6}$$

where

$$C = \sum_{i=1}^{k} b_i \quad \text{and} \quad D = \sum_{i=0}^{k-1} b_i. \tag{3.1.7}$$

First assume that (i) is satisfied. Then for $x \leq y \leq \bar{x}$, the following inequalities hold:

$$
\begin{aligned}
1 + b_0 y + Cx &\geq 1 + b_0 x + Cx = 1 + Bx, \\
1 + D\bar{x} + b_r y &\geq 1 + D\bar{x} + b_r x \geq 1 + Bx, \\
1 + b_0\bar{x} + Cx &\geq 1 + b_0 x + Cx = 1 + Bx, \\
a + by &\leq a + b\bar{x}.
\end{aligned}
$$

Hence

$$G(x, y) \leq \frac{\bar{x}\left(b + \dfrac{a}{\bar{x}}\right)^{k+1}}{(1 + Bx)^{k+1}}$$

and so for $x \leq \bar{x}$,

$$F(x) = \max_{x \leq y \leq \bar{x}} G(x,y) \leq \varphi_{k+1}(x)$$

where

$$\varphi_p(x) = \gamma\left(\frac{\alpha}{1 + \beta x}\right)^p \quad \text{for} \quad x \geq 0 \tag{3.1.8}$$

with

$$\alpha = b + \frac{a}{\bar{x}} > 1, \; \beta = B, \quad \text{and} \quad \gamma = \bar{x}.$$

Here condition (i) implies that

$$(\alpha - 1)(p - 1) \leq 1 \tag{3.1.9}$$

holds with $p = k + 1$. Similarly, for $x \geq \bar{x}$ we find

$$F(x) = \min_{\bar{x} \leq y \leq x} G(x,y) \geq \varphi_{k+1}(x).$$

By Lemma 1.6.6 the unique positive solution of the equation

$$\varphi_p(\varphi_p(x)) = x$$

is $x = \bar{x}$, and so by Lemma 1.6.4

$$F(F(x)) = x$$

also has the unique positive solution $x = \bar{x}$. Hence by Theorem 2.3.1, $\bar{x}$ is a global attractor of all positive solutions of Eq.(3.1.1).

Next assume that (ii) holds. It is now easy to see that the function f satisfies condition (a) of Theorem 2.3.2, and the function F_1 is given by

$$F_1(x) = \frac{\bar{x}\left(b + \frac{a}{\bar{x}}\right)^{k+1}}{(1 + b_0\bar{x} + Cx)^{k+1}} = \varphi_{k+1}(x)$$

with

$$\alpha = \frac{b + \frac{a}{\bar{x}}}{1 + b_0\bar{x}} > 1, \; \beta = \frac{C}{1 + b_0\bar{x}}, \quad \text{and} \quad \gamma = \bar{x}.$$

Also the first condition in (ii) implies that (3.1.9) holds with $p = k + 1$. By Lemma 1.6.6 we see that the equation $F_1(F_1(x)) = x$ has the unique positive solution $\bar{x}$, and so by Theorem 2.3.2, $\bar{x}$ is a global attractor of all positive solutions of Eq.(3.1.1).

Finally, assume that (iii) holds. Then the function $(a + by)/(1 + D\bar{x} + b_ky)$ is increasing in y and, for $x \leq y \leq \bar{x}$, we have

$$\frac{a + by}{1 + D\bar{x} + b_ky} \leq \frac{a + b\bar{x}}{1 + D\bar{x} + b_k\bar{x}} = \bar{x}$$
$$1 + b_0y + Cx \geq 1 + Bx,$$
$$z1 + b_0\bar{x} + Cx \geq 1 + Bx,$$

and

$$G(x,y) \le \bar{x} \left(\frac{b + \dfrac{a}{\bar{x}}}{1 + Bx} \right)^k .$$

Then for $x \le \bar{x}$,

$$F(x) = \max_{x \le y \le \bar{x}} G(x,y) \le \varphi_k(x)$$

where

$$\alpha = b + \frac{a}{\bar{x}} > 1, \ \beta = B, \ \text{and} \ \gamma = \bar{x}.$$

Also condition (iii) implies that (3.1.9) holds with $p = k$. Similarly for $x > \bar{x}$, we have

$$F(x) = \min_{\bar{x} \le y \le x} G(x,y) \ge \varphi_k(x)$$

and the result follows by an argument similar to the one given above. The proof is complete. □

Research Project 3.1.1 *Extend the results of this section to equations where some of the coefficients are negative.*

Research Project 3.1.2 *Extend the results of this section to equations with complex coefficients.*

3.2 The Rational Recursive Sequence $$x_{n+1} = (a + \sum_{i=0}^{k} a_i x_{n-i})/(b + \sum_{i=0}^{k} b_i x_{n-i})$$

In this section we investigate the global attractivity of all positive solutions of the difference equation

$$x_{n+1} = \frac{a + \sum\limits_{i=0}^{k} a_i x_{n-i}}{b + \sum\limits_{i=0}^{k} b_i x_{n-i}}, \ n = 0, 1, \dots \tag{3.2.1}$$

where k is a nonnegative integer

$$\left. \begin{array}{l} a_0, \dots, a_k, b_0, \dots, b_k \in [0, \infty), a, b \in (0, \infty) \\[2mm] \sum\limits_{i=0}^{k} a_i = 1, \ \text{and} \ B = \sum\limits_{i=0}^{k} b_i > 0 \end{array} \right\} \tag{3.2.2}$$

and where the initial conditions $x_{-k}, \ldots, x_0$ are arbitrary positive numbers.

Eq.(3.2.1) has the unique positive equilibrium

$$\bar{x} = \frac{1 - b + \sqrt{(1 - b)^2 + 4aB}}{2B}.$$

Lemma 3.2.1 *Assume that (3.2.2) holds. Let $\{x_n\}$ be a solution of Eq.(3.2.1) such that for some $n_0 \geq 0$,*

$$\text{either} \quad x_n \geq \bar{x} \quad \text{for} \quad n \geq n_0 \tag{3.2.3}$$

$$\tag{3.2.4}$$

$$\text{or} \quad x_n \leq \bar{x} \quad \text{for} \quad n \geq n_0. \tag{3.2.5}$$

Then

$$\lim_{n \to \infty} x_n = \bar{x}. \tag{3.2.6}$$

Proof. Assume that (3.2.3) holds. The case where (3.2.5) holds is similar and will be omitted. For $n \geq n_0 + k$,

$$x_{n+1} = \frac{a + \sum_{i=0}^{k} a_i x_{n-i}}{b + \sum_{i=0}^{k} b_i x_{n-i}} = \left[\sum_{i=0}^{k} a_i x_{n-i} \right] \left[\frac{1 + \frac{a}{\sum_{i=0}^{k} a_i x_{n-i}}}{b + \sum_{i=0}^{k} b_i x_{n-i}} \right]$$

$$\leq \left[\sum_{i=0}^{k} a_i x_{n-i} \right] \frac{1 + \frac{a}{\bar{x}}}{b + B\bar{x}} = \sum_{i=0}^{k} a_i x_{n-i}$$

and so

$$x_{n+1} \leq \max_{0 \leq i \leq k} \{x_{n-i}\} \quad \text{for} \quad n \geq n_0 + k. \tag{3.2.7}$$

Set

$$y_n = \max_{0 \leq i \leq k} \{x_{n-i}\} \quad \text{for} \quad n \geq n_0 + k. \tag{3.2.8}$$

Then clearly

$$y_n \geq x_{n+1} \geq \bar{x} \quad \text{for} \quad n \geq n_0 + k. \tag{3.2.9}$$

Next we claim that

$$y_{n+1} \leq y_n \quad \text{for} \quad n \geq n_0 + k. \tag{3.2.10}$$

Indeed

$$y_{n+1} = \max_{0 \leq i \leq k} \{x_{n+1-i}\} = \max\{x_{n+1}, \max_{0 \leq i \leq k-1} \{x_{n-i}\}\}$$

$$\leq \max\{x_{n+1}, y_n\} = y_n.$$

From (3.2.9) and (3.2.10) it follows that the sequence $\{y_n\}$ is convergent and that

$$y \equiv \lim_{n \to \infty} y_n \geq \bar{x}. \tag{3.2.11}$$

Furthermore

$$x_{n+1} \leq \frac{a + \sum_{i=0}^{k} a_i x_{n-i}}{b + B\bar{x}} \leq \frac{a + y_n}{b + B\bar{x}}.$$

From this and by using (3.2.10) we find,

$$x_{n+i} \leq \frac{a + y_{n+i-1}}{b + B\bar{x}} \leq \frac{a + y_n}{b + B\bar{x}} \quad \text{for} \quad i = 1, \ldots, k+1.$$

Then

$$y_{n+k+1} = \max_{1 \leq i \leq k+1} \{x_{n+i}\} \leq \frac{a + y_n}{b + B\bar{x}} \tag{3.2.12}$$

and by letting $n \to \infty$ we obtain

$$y \leq \frac{a + y}{b + B\bar{x}} = \frac{\bar{x}}{a + \bar{x}}(a + y).$$

Hence $y \leq \bar{x}$ and so in view of (3.2.11), $y = \bar{x}$. The proof is complete. $\square$

A sequence is said to **persist** if it is bounded away from zero by a positive constant.
The following theorem gives sufficient conditions for the boundedness and persistence of all positive solutions of Eq.(3.2.1).

Theorem 3.2.1 *Assume that (3.2.2) holds. Let $\{x_n\}$ be a positive solution of Eq. (3.2.1) and assume that one of the following two conditions is satisfied:*

(a) $b > 1$;

(b) $b \leq 1$ and for each $i = 0, \ldots, k$ for which a_i is positive, b_i is also positive.

Then there exist positive constants C and D such that

$$C \leq x_n \leq D \quad \text{for} \quad n = 0, 1, \ldots. \tag{3.2.13}$$

Proof. From Eq.(3.2.1) we have

$$x_{n+1} \leq \frac{a}{b} + \frac{1}{b} \sum_{i=0}^{k} a_i x_{n-i}, \quad n = 0, 1, \ldots. \tag{3.2.14}$$

Consider the linear difference equation

$$y_{n+1} = \frac{a}{b} + \frac{1}{b} \sum_{i=0}^{k} a_i y_{n-i}, \quad n = 0, 1, \dots \qquad (3.2.15)$$

with initial conditions

$$y_i = x_i \geq 0 \quad \text{for} \quad i = -k, \dots, 0.$$

It follows by induction that

$$x_n \leq y_n \quad \text{for} \quad n = -k, \dots. \qquad (3.2.16)$$

First assume that $b > 1$. Observe that $a/(b-1)$ is a particular solution of Eq.(3.2.15), while in view of Remark 1.3.1, every solution of the homogeneous equation which is associated with Eq.(3.2.15) tends to zero as $n \to \infty$. Hence

$$\lim_{n \to \infty} y_n = \frac{a}{b-1}.$$

From this and (3.2.16) it follows that the sequence $\{x_n\}$ is bounded from above by a positive constant, say D. That is,

$$x_n \leq D \quad \text{for} \quad n = 0, 1, \dots.$$

Set

$$C = \frac{a}{b + BD}$$

and observe that

$$x_{n+1} = \frac{a + \displaystyle\sum_{i=0}^{k} a_i x_{n-i}}{b + \displaystyle\sum_{i=0}^{k} b_i x_{n-i}} \geq \frac{a}{b + BD} = C$$

which completes the proof when (a) holds.

Next consider the case where (b) holds. It suffices to show that $\{x_n\}$ is bounded from above by some positive constant. For the sake of contradiction, assume that $\{x_n\}$ is unbounded. Then there exists a subsequence $\{x_{n_j}\}$ such that

$$\lim_{j \to \infty} n_j = \infty, \quad \lim_{j \to \infty} x_{1+n_j} = \infty$$

and

$$x_{1+n_j} = \max\{x_n : -k \leq n \leq 1 + n_j\} \quad \text{for} \quad j = 0, 1, \dots.$$

Then (3.2.14) implies that

$$\lim_{j \to \infty} \sum_{i=0}^{k} a_i x_{n_j - i} = \infty. \tag{3.2.17}$$

As $\sum_{i=0}^{k} a_i = 1$, it follows that

$$\sum_{i=0}^{k} a_i x_{n_j - i} \le x_{1 + n_j},$$

and so

$$0 \le x_{1 + n_j} - \sum_{i=0}^{k} a_i x_{n_j - i} = \frac{a + (\sum_{i=0}^{k} a_i x_{n_j - i})(1 - b - \sum_{i=0}^{k} b_i x_{n_j - i})}{b + \sum_{i=0}^{k} b_i x_{n_j - i}}.$$

From this and (3.2.17) it follows that

$$\sum_{i=0}^{k} b_i x_{n_j - i} \le 1 - b.$$

Then for every $i = 0, \ldots, k$ for which b_i is positive, the subsequence $\{x_{n_j - i}\}$ is bounded which implies that the sequence $\{\sum_{i=0}^{k} a_i x_{n_j - i}\}$ is also bounded. This contradicts (3.2.17) and the proof is complete. $\square$

Theorem 3.2.2 *Assume that (3.2.2) holds and that $b > 1$. Then the positive equilibrium $\bar{x}$ of Eq.(3.2.1) is globally asymptotically stable.*

Proof. The linearized equation associated with Eq.(3.2.1) is

$$z_{n+1} + \sum_{i=0}^{k} \frac{b_i \bar{x} - a_i}{b + B\bar{x}} z_{n-i} = 0, \quad n = 0, 1, \ldots.$$

As $b > 1$,

$$\sum_{i=0}^{k} \frac{b_i \bar{x} - a_i}{b + B\bar{x}} \le \frac{1 + B\bar{x}}{b + B\bar{x}} < 1$$

and so by Remark 1.2.1 we see that $\bar{x}$ is locally asymptotically stable.

To complete the proof it remains to show that $\bar{x}$ is a global attractor of all positive solutions of Eq.(3.2.1). Set

$$\lambda = \liminf_{n \to \infty} x_n \quad \text{and} \quad \Lambda = \limsup_{n \to \infty} x_n$$

which by Theorem 3.2.1 exist and are positive numbers. Then from Eq.(3.2.1) we see that

$$\Lambda \leq \frac{a + \Lambda}{b + B\lambda} \quad \text{and} \quad \lambda \geq \frac{a + \lambda}{b + B\Lambda}.$$

Hence

$$a + (1 - b)\lambda \leq B\lambda\Lambda \leq a + (1 - b)\Lambda$$

from which it follows that $\lambda = \Lambda$. The proof is complete. □

Research Project 3.2.1 *Extend the results of this section to equations where some of the coefficients are negative.*

Research Project 3.2.2 *Extend the results of this section to equations with complex coefficients.*

3.3 Global Stability of $x_{n+1} = 1/\sum_{i=0}^{k} b_i x_{n-i}$

In this section we investigate the global stability of all positive solutions of the difference equation

$$x_{n+1} = \frac{1}{\displaystyle\sum_{i=0}^{k} b_i x_{n-i}}, \quad n = 0, 1, \ldots \tag{3.3.1}$$

where

$$b_0, \ldots, b_k \in (0, \infty) \quad \text{and} \quad k \in \{1, 2, \ldots\}. \tag{3.3.2}$$

Eq.(3.3.1) has the unique positive equilibrium

$$\bar{x} = \frac{1}{\sqrt{B}} \tag{3.3.3}$$

where

$$B = \sum_{i=0}^{k} b_i. \tag{3.3.4}$$

The following theorem establishes a global asymptotic stability result for the positive solutions of Eq.(3.3.1).

Theorem 3.3.1 *Assume that (3.3.2) holds. Then $\bar{x}$ is a globally asymptotically stable equilibrium of all solutions of Eq.(3.3.1) with positive initial conditions.*

Proof. Choose $M > \bar{x}$ in such a way that

$$x_{-k}, \ldots, x_0 \in \left(\frac{1}{BM}, M \right).$$

It is easy to see that $\bar{x} \in (\frac{1}{BM}, M)$. Then it follows by induction that

$$\frac{1}{BM} < x_n < M \quad \text{for} \quad n \geq -k. \tag{3.3.5}$$

Let

$$\lambda = \liminf_{n \to \infty} x_n \quad \text{and} \quad \mu = \limsup_{n \to \infty} x_n.$$

Since μ is finite, for any positive ϵ there exists an integer n_0 such that

$$x_n < \mu + \epsilon \quad \text{for} \quad n \geq n_0.$$

Then

$$x_{n+1} = \frac{1}{\displaystyle\sum_{i=0}^{k} b_i x_{n-i}} > \frac{1}{B(\mu + \epsilon)} \quad \text{for} \quad n \geq n_0 + k$$

from which it follows that $\lambda \geq 1/B\mu$. Similarly we can show that $\mu \leq 1/B\lambda$. Hence

$$\lambda\mu = \frac{1}{B}.$$

Let $\{n_i\}$ be an infinite increasing sequence of positive numbers such that

$$\lim_{i \to \infty} x_{n_i+1} = \mu.$$

By taking subsequences, if necessary, we also assume that the following $(k + 2)$ subsequences $\{x_{n_i}\}$, $\{x_{n_i-1}\}$, $\ldots$, $\{x_{n_i-k-1}\}$ approach L_0, L_1, $\ldots$, L_{k+1}, respectively. Since

$$\sum_{i=0}^{k} b_i x_{n-i} = \frac{1}{x_{n+1}} \quad \text{and} \quad \sum_{i=0}^{k} b_i x_{n-i-1} = \frac{1}{x_n}$$

it follows that

$$\sum_{i=0}^{k} b_i L_i = \frac{1}{\mu} = B\lambda \quad \text{and} \quad \sum_{i=0}^{k} b_i L_{i+1} = \frac{1}{L_0}.$$

Clearly

$$L_0, \ldots, L_{k+1} \in [\lambda, \mu]$$

and so
$$L_0 = \cdots = L_k = \lambda.$$

Then
$$\sum_{i=0}^{k} b_i L_{i+1} = \frac{1}{\lambda} = b\mu$$

and so
$$L_1 = \cdots = L_{k+1} = \mu.$$

Hence $\lambda = \mu = \bar{x}$ which implies that $\bar{x}$ is a global attractor of all positive solutions of Eq.(3.3.1).

To complete the proof it remains to show that $\bar{x}$ is locally asymptotically stable. To this end, let $\epsilon \in (0, \bar{x})$ be arbitrary. Set

$$M(\epsilon) = \min\{\bar{x} + \epsilon, \frac{1}{B(\bar{x} - \epsilon)}\}$$

and

$$\delta(\epsilon) = \min\{M(\epsilon) - \bar{x}, \bar{x} - \frac{1}{BM(\epsilon)}\}.$$

It is easy to see that

$$(\bar{x} - \delta, \bar{x} + \delta) \subseteq (\frac{1}{BM}, M) \subseteq (\bar{x} - \epsilon, \bar{x} + \epsilon). \qquad (3.3.6)$$

Let $x_{-k}, \ldots, x_0 \in (\bar{x} - \delta, \bar{x} + \delta)$. Then from (3.3.5) and (3.3.6) it follows that $x_n \in (\bar{x} - \epsilon, \bar{x} + \epsilon)$ for $n \geq -k$ which shows that $\bar{x}$ is locally asymptotically stable. The proof is complete. $\qquad \square$

Remark 3.3.1 It is easy to see that all solutions of Eq.(3.3.1) are oscillatory and that each semicycle has at most $(k + 1)$ terms.

Remark 3.3.2 The assumption $k \geq 1$ is essential. Indeed when $k = 0$, Eq.(3.3.1) becomes

$$x_{n+1} = \frac{1}{b_0 x_n}, \quad n = 0, 1, \ldots$$

and all solutions are periodic with period 2.

Research Project 3.3.1 *Extend the results of this section to equations where some of the coefficients are negative.*

Research Project 3.3.2 *Extend the results of this section to equations with complex coefficients.*

3.4 Global Stability of $x_{n+1} = (a + bx_n)/(A + x_{n-k})$

In this section we investigate the recursive sequence

$$x_{n+1} = \frac{a + bx_n}{A + x_{n-k}}, \quad n = 0, 1, \ldots \tag{3.4.1}$$

where

$$a, b \in [0, \infty) \text{ with } a + b > 0, \quad A \in (0, \infty), \text{ and } k \in \{1, 2, \ldots\}. \tag{3.4.2}$$

Equation (3.4.1) has a positive equilibrium $\bar{x}$ provided that

$$\text{either } a > 0 \text{ or } a = 0 \text{ and } b > A. \tag{3.4.3}$$

When (3.4.3) holds, Eq.(3.4.1) has the unique positive equilibrium

$$\bar{x} = \frac{b - A + \sqrt{(b - A)^2 + 4a}}{2}. \tag{3.4.4}$$

The linearized equation of Eq.(3.4.1) about the positive equilibrium $\bar{x}$ is

$$y_{n+1} - \frac{b}{A + \bar{x}} y_n + \frac{\bar{x}}{A + \bar{x}} y_{n-k} = 0, \; n = 0, 1, \ldots. \tag{3.4.5}$$

Lemma 3.4.1 *Assume that (3.4.2) and (3.4.3) hold. Then in each of the following four cases, the positive equilibrium $\bar{x}$ of Eq.(3.4.1) is locally asymptotically stable:*

(a) $b = 0$;

(b) $a = 0$ and $b > A > b\left(1 - 2\cos\dfrac{k\pi}{2k + 1}\right)$;

(c) $k = 1$ and $a > 0$;

(d) $k \geq 2$ and $A > b$;

Proof. (a) The characteristic equation of the linearized equation (3.4.5) is

$$\lambda^{k+1} + \frac{\bar{x}}{A + \bar{x}} = 0$$

whose roots lie inside the unit disk $|\lambda| < 1$.
(b) The linearized equation (3.4.5) takes the form

$$y_{n+1} - y_n + \frac{b - A}{A} y_{n-k} = 0, \; n = 0, 1, \ldots \tag{3.4.6}$$

and the result follows by applying Theorem 1.3.6.

(c) The result is a simple application of Theorem 1.3.4.

(d) This result is a simple application of Theorem 1.3.7. $\square$

The following result establishes the boundedness and persistence of all positive solutions of Eq.(3.4.1).

Theorem 3.4.1 *Assume that (3.4.2) holds and let $\{x_n\}$ be a solution of Eq.(3.4.1) with positive initial conditions $x_{-k}, \ldots, x_0$. Then the following statements are true:*

(a) Assume that (3.4.3) holds. Then Eq.(3.4.1) is permanent.

(b) Assume that $a = 0$ and $0 < b \le A$. Then $\{x_n\}$ is decreasing and

$$\lim_{n \to \infty} x_n = 0.$$

Proof. (a) First consider the case where $b > 0$. Then Eq.(3.4.1) can be written in the form

$$x_{n+1} = \frac{\dfrac{a}{A} + \dfrac{b}{A}x_n}{1 + \dfrac{1}{A}x_{n-k}}, \quad n = 0, 1, \ldots$$

which is a special case of Eq.(3.1.1). The result follows by applying Theorem 3.1.1(a).

Next consider the case where $b = 0$. Then $a > 0$ and

$$x_{n+1} = \frac{a}{A + x_{n-k}} \le \frac{a}{A} \quad \text{for} \quad n = 0, 1, \ldots.$$

From this it follows that

$$x_{n+1} \ge \frac{a}{A + a/A} = \frac{aA}{a^2 + a} \quad \text{for} \quad n = 0, 1, \ldots$$

and the proof of (a) is complete.

(b) The result follows by applying Theorem 3.1.1(b). $\square$

The following result describes among other things the oscillatory character of all positive solutions of Eq.(3.4.1).

Theorem 3.4.2 *Assume that (3.4.2) and (3.4.3) hold and let $\{x_n\}$ be a positive solution of Eq.(3.4.1). Then the following statements are true:*

(a) Suppose $b > 0$, and assume that for some $n_0 \ge 0$,

$$\text{either} \quad x_n \ge \bar{x} \quad \text{for} \quad n \ge n_0 \quad \text{or} \quad x_n \le \bar{x} \quad \text{for} \quad n \ge n_0.$$

Then for $n \ge n_0 + k$, $\{x_n\}$ is monotonic and

$$\lim_{n \to \infty} x_n = \bar{x}.$$

(b) Suppose b > 0, and assume that $\{x_n\}$ is a positive solution of Eq. (3.4.1) which oscillates about $\bar{x}$. Then the extreme point in every semicycle occurs in one of the first $(k+1)$ terms.

(c) Suppose b > 0, and suppose that

$$\frac{\bar{x}(A+\bar{x})^k}{b^{k+1}} > \frac{k^k}{(k+1)^{k+1}}. \tag{3.4.7}$$

Then every solution of Eq.(3.4.1) oscillates about the positive equilibrium $\bar{x}$.

(d) Assume that b > 0 and a > bA. Then every nontrivial positive solution of Eq.(3.4.1) is strictly oscillatory about the positive equilibrium $\bar{x}$. Furthermore, a semicycle of such a solution has at most $(2k+1)$ terms.

(e) Assume that b = 0. Then each positive solution of Eq.(3.4.1) oscillates about $\bar{x}$ and each semicycle has at most $(k+1)$ terms.

Proof. Parts (a) and (b) are direct consequences of Lemmas 2.2.1 and 2.2.2, respectively.

(c) Assume for the sake of contradiction that Eq.(3.4.1) has a positive solution $\{x_n\}$ which is not oscillatory about $\bar{x}$. Then for some $n_0 \geq 0$, either

$$x_n \geq \bar{x} \quad \text{for} \quad n \geq n_0 \tag{3.4.8}$$

or

$$x_n \leq \bar{x} \quad \text{for} \quad n \geq n_0 \tag{3.4.9}$$

and in either case

$$\lim_{n\to\infty} x_n = \bar{x}. \tag{3.4.10}$$

We will assume that (3.4.8) holds. The case where (3.4.9) holds is similar and will be omitted. Set

$$x_n = z_n + \bar{x} \quad \text{for} \quad n \geq -k.$$

Then Eq.(3.4.1) yields,

$$z_{n+1} - z_n + \frac{A+\bar{x}-b}{A+x_{n-k}}z_n + \frac{x_n}{A+x_{n-k}}z_{n-k} = 0$$

which shows that $\{z_n\}$ is an eventually positive solution of the difference equation

$$u_{n+1} - u_n + P(n)u_n + Q(n)u_{n-k} = 0$$

where

$$P(n) = \frac{A+\bar{x}-b}{A+x_{n-k}} \quad \text{and} \quad Q(n) = \frac{x_n}{A+x_{n-k}}.$$

Note from (3.4.4) that $\bar{x} > b - A$ and so because of (3.4.10) the limits

$$\lim_{n \to \infty} P(n) = \frac{A + \bar{x} - b}{A + \bar{x}} \quad \text{and} \quad \lim_{n \to \infty} Q(n) = \frac{\bar{x}}{A + \bar{x}}$$

exist and are positive numbers. By applying Theorem 1.2.2(a) we see that the equation

$$\lambda - 1 + \frac{A + \bar{x} - b}{A + \bar{x}} + \frac{\bar{x}}{A + \bar{x}} \lambda^{-k} = 0,$$

or equivalently,

$$F(\lambda) \equiv \lambda^{k+1} - \frac{b}{A + \bar{x}} \lambda^k + \frac{\bar{x}}{A + \bar{x}} = 0$$

has a positive root. By computing the extreme of the polynomial F we see that

$$\min_{\lambda > 0} F(\lambda) > 0$$

if and only if (3.4.7) holds. Hence the function F cannot have a positive zero. This is a contradiction and the proof is complete.

(d) Assume for the sake of contradiction that Eq.(3.4.1) has a nontrivial positive solution $\{x_n\}$ which is not strictly oscillatory about $\bar{x}$. Then for some $n_0 \geq 0$, either (3.4.8) or (3.4.9) holds. Suppose that (3.4.8) holds. The case where (3.4.9) holds is similar and will be omitted. By part (a), $\{x_n\}$ is decreasing for $n \geq n_0 + k$. Also because the solution is nontrivial,

$$x_{n_0+k} > \bar{x}.$$

Observe that because $a > bA$, the function $g(x) = \dfrac{a + bx}{A + x}$ is strictly decreasing. Thus

$$\bar{x} = g(\bar{x}) > g(x_{n_0+k}) = \frac{a + bx_{n_0+k}}{A + x_{n_0+k}} \geq \frac{a + bx_{n_0+2k}}{A + x_{n_0+k}} = x_{n_0+2k+1}$$

which is a contradiction. Therefore every positive solution of Eq.(3.4.1) is strictly oscillatory. The above analysis also shows that no semicycle may contain more than $(2k + 1)$ terms. The proof is complete.

(e) When $b = 0$,

$$x_{n+1} - \bar{x} = \frac{a}{A + x_{n-k}} - \bar{x} = \bar{x} \frac{\bar{x} - x_{n-k}}{A + x_{n-k}}$$

from which the result follows. The proof is complete.   □

Remark 3.4.1 When $a = 0$ and $b > 0$, (3.4.7) is also necessary for the oscillation of all solutions of Eq.(3.4.1). Indeed, the change of variables

$$x_n = \bar{x} e^{v_n}$$

transforms Eq.(3.4.1) to the difference equation

$$v_{n+1} - v_n + \frac{b-A}{b} f(v_{n-k}) = 0, \ n = 0, 1, \ldots . \qquad (3.4.11)$$

By applying Theorem 1.2.3 to Eq.(3.4.11) and then by using Corollary 1.2.1, we see that all solutions of Eq.(3.4.1), when $a = 0$, are oscillatory if and only if (3.4.7) holds.

The following result deals with the global attractivity of the positive equilibrium of Eq.(3.4.1).

Theorem 3.4.3 *Assume that (3.4.2) and (3.4.3) hold. Then the positive equilibrium $\bar{x}$ of Eq.(3.4.1) is a global attractor of all positive solutions provided that one of the following six conditions is satisfied:*

(a) $a > 0$ and $A > b > 0$;

(b) $a > 0$ and $b = 0$;

(c) $b > 0$, $k \geq 2$, $Ab < a$, and $\bar{x}k \leq A$;

(d) $b > 0$, $k \geq 2$, $a \leq Ab$, and $\bar{x}(k-1) \leq A$;

(e) $b > 0$, $k = 1$, and $a \leq Ab$;

(f) $b > 0$, $k = 1$, and $Ab \leq a \leq 2Ab + 2A^2$.

Proof. (a) Equation (3.4.1) can be written in the form

$$x_{n+1} = \frac{\dfrac{a}{b} + x_n}{\dfrac{A}{b} + \dfrac{1}{b} x_{n-k}}, \ n = 0, 1, \ldots$$

which is a special case of Eq.(3.2.1). The global attractivity of $\bar{x}$ now follows by applying Theorem 3.2.1(a).
(b) Consider the function $F(u) = a/(A+u)$. Clearly F is continuous on $(0, \infty)$, F is decreasing, and the system

$$U = \frac{a}{A+L} \quad \text{and} \quad L = \frac{a}{A+U}$$

has the unique solution $U = L = \bar{x}$. The result follows by applying Theorem 2.4.1 with $\alpha = 0$.
(c) and (d) follow from Theorem 3.1.1(c).

(e) The result follows by applying Theorem 2.1.1.

(f) We will apply Theorem 2.3.1. It is easy to see that the function

$$f(u_0, u_1) = \frac{a + \dfrac{b}{u_0}}{A + u_1}$$

satisfies the hypotheses (H$_1$) - (H$_4$) of Section 2.2. Furthermore the function G which is defined by (2.3.1) takes the form

$$G(x, y) = \frac{a + by}{A + y} \frac{a + \frac{b}{x}}{A + x}. \tag{3.4.12}$$

Next we will construct the function F defined by (2.3.2). Since

$$\frac{d}{dy}\left(\frac{a + by}{A + y}\right) = \frac{bA - a}{(A + y)^2}$$

the following two cases are possible:

Case (i): $bA \geq a$. Then the function $(a + by)/(A + y)$ is increasing and

$$\max_{x \leq y \leq \bar{x}} \frac{a + by}{A + y} = \frac{a + b\bar{x}}{A + \bar{x}} = \bar{x}.$$

Also

$$\min_{\bar{x} \leq y \leq x} \frac{a + by}{A + y} = \frac{a + b\bar{x}}{A + \bar{x}} = \bar{x}.$$

Hence the function F is given by

$$F(x) = \frac{a + b\bar{x}}{A + x}.$$

It is easy to see that the only positive solution of the equation $F^2(x) = x$ is $x = \bar{x}$. The global attractivity of $\bar{x}$ now follows by applying Theorem 2.3.1.

Case (ii): $bA < a$. Then the function $(a + by)/(A + y)$ is decreasing and

$$\max_{x \leq y \leq \bar{x}} \frac{a + by}{A + y} = \frac{a + bx}{A + x}.$$

Also

$$\min_{\bar{x} \leq y \leq x} \frac{a + by}{A + y} = \frac{a + bx}{A + x}.$$

Hence the function F is given by

$$F(x) = \frac{(A + \bar{x})(a + bx)}{(A + x)^2}.$$

To complete the proof it remains to show that the only solution of the system

$$y = \frac{(A+\bar{x})(a+bx)}{(A+x)^2} \quad \text{and} \quad x = \frac{(A+\bar{x})(a+by)}{(A+y)^2} \qquad (3.4.13)$$

is $x = y = \bar{x}$. For the sake of contradiction, assume that there exists a solution $\{x, y\}$ of the system (3.4.13) such that $0 < x < \bar{x} < y$. Then it is easy to see that $x = F(y)$ and $y = F(x)$. From (3.4.13) it follows that

$$y(A+x)^2 = (A+\bar{x})(a+bx) \quad \text{and} \quad x(A+y)^2 = (A+\bar{x})(a+by)$$

and so

$$xy = A^2 + (A+\bar{x})b. \qquad (3.4.14)$$

Furthermore, from (3.4.13) and (3.4.14) we see that x and y satisfy the equation

$$\frac{A^2 + (A+\bar{x})b}{u} = \frac{(A+\bar{x})(a+bu)}{(A+u)^2}$$

which is equivalent to

$$h(u) \equiv A^2u^2 + (2A^3 + (A+\bar{x})(2Ab - a))u + A^2(A^2 + (A+\bar{x})b) = 0.$$

Since $h(0) > 0$, $h(\infty) > 0$, and $h(\bar{x}) = (A+\bar{x})^3(A+b-\bar{x})$, it is easy to see that the condition $a \leq 2Ab + 2A^2$ is equivalent to $\bar{x} \leq A + b$. Then $h(\bar{x}) \geq 0$, and this implies that function $h(u)$ has no zeros x and y such that $0 < x < \bar{x} < y$. This is a contradiction and the proof is complete. □

From Lemma 3.4.1 and Theorem 3.4.3 we obtain the following result.

Corollary 3.4.1 *Assume that (3.4.2) and (3.4.3) hold. Then the positive equilibrium $\bar{x}$ of Eq.(3.4.1) is globally asymptotically stable provided that one of the following five conditions is satisfied:*

(a) $a > 0$ and $b = 0$;

(b) $a > 0$, $k \geq 2$, and $A > b > 0$;

(c) $a = 0$ and $b > A \geq b\dfrac{k-1}{k}$;

(d) $a > 0$, $b > 0$, $k = 1$, and $a \leq Ab$;

(e) $a > 0$, $b > 0$, $k = 1$, and $Ab \leq a \leq 2Ab + 2A^2$.

Research Project 3.4.1 *Extend the results of this section to equations where some of the coefficients are negative.*

Research Project 3.4.2 *Extend the results of this section to equations with complex coefficients.*

Research Project 3.4.3 *Assume that the sequences $\{a_n\}$, $\{b_n\}$, and $\{A_n\}$ are periodic. Investigate the behavior of solutions of the rational recursive sequence*

$$x_{n+1} = \frac{a_n + b_n x_n}{A_n + x_{n-1}}, \ n = 0, 1, \ldots.$$

3.5   Notes

The results in Sections 3.1 and 3.2 are from Kocic and Ladas [6]. Theorem 3.3.1 is extracted from a solution given in Borwein [1]. The results in Section 3.4 are extracted from Kocic and Ladas [3,6], Kocic, Ladas and Rodrigues [1] and Kuruklis and Ladas [1].

The reader should note that our work in this chapter has just scratched the surface of rational recursive sequences. Computer observations and some analytical work (see Camouzis, Ladas and Rodrigues [1]) show, for example, that the solutions of the rational recursive sequence

$$x_{n+1} = \frac{a + b x_n^2}{c + x_{n-k}^2}, \ n = 0, 1, \ldots$$

where

$$k \in \{0, 1, \ldots\} \quad \text{and} \quad a, b, c, \in [0, \infty) \quad \text{with} \quad a + b > 0$$

have fascinating properties and also a great deal of complication which is not present in Eq.(3.4.1).

Chapter 4

Applications

In this chapter we inverstigate the global asymptotic stability and the oscillatory character of some discrete models, as well as some discrete analogues of continuous models taken from Mathematical Biology and Physics.

4.1 A Discrete Delay Logistic Model

The discrete delay logistic model

$$N_{t+1} = \frac{\alpha N_t}{1 + \beta N_{t-k}}, \ t \geq 0 \tag{4.1.1}$$

where

$$\alpha \in (1, \infty), \beta \in (0, \infty) \quad \text{and} \quad k \in \{0, 1, \ldots\} \tag{4.1.2}$$

was proposed by Pielou in her books [1, p.22] and [2, p.79] as a discrete analogue of the delay logistic equation

$$\dot{N}(t) = rN(t)[1 - \frac{N(t - \tau)}{P}], \ t \geq 0. \tag{4.1.3}$$

One arrives at Eq.(4.1.3) from the logistic equation

$$\dot{N}(t) = rN(t)[1 - \frac{N(t)}{P}], \ t \geq 0 \tag{4.1.4}$$

by assuming that there is a delay τ in the response of the growth rate per individual to density changes.

Eq.(4.1.3) is a prototype in modelling the dynamics of single-species. Here $N(t)$ denotes the density or biomass of the population at time t, the positive constant r is the growth rate of the population and the positive constant P is the carrying capacity of the environment. The term $1 - N(t - \tau)/P$ denotes a feedback mechanism which takes τ units of time to respond to changes in the size of the population.

Pielou arrived at her model as follows: She observed that the solution

$$N(t) = \frac{P}{1 + (\frac{P}{N(0)} - 1)e^{-rt}}$$

of Eq.(4.1.4) satisfies the functional relationship

$$N(t + 1) = \frac{\alpha N(t)}{1 + \beta N(t)}, \ t \geq 0 \tag{4.1.5}$$

where

$$\alpha = e^r > 1 \quad \text{and} \quad \beta = \frac{e^r - 1}{P} > 0.$$

Finally from (4.1.5) she was led to Eq.(4.1.1) by assuming, as in the continuous case, that there should be a delay k in the response of the growth rate per individual to density changes.

Pielou's interest in Eq.(4.1.1) was to show that "the tendency to oscillate is a property of the populations themselves and is independent of any extrinsic factors". That is, population sizes oscillate "even though the environment remains constant". According to Pielou, "oscillations can be set up in a population if its growth rate is governed by a density dependent mechanism and if there is a delay in the response of the growth rate to density changes. When this happens the size of the population alternately overshoots and undershoots its equilibrium level".

The blow-fly *Lucilia cuprina* which was studied by Nicholson [1] is an example of a laboratory population which behaves in the manner described above.

For simplicity we will write Eq.(4.1.1) in the form

$$x_{n+1} = \frac{\alpha x_n}{1 + \beta x_{n-k}}, \ n = 0, 1, \dots. \tag{4.1.6}$$

Now observe that this equation is of the form of Eq.(3.4.1) with

$$a = 0, \quad b = \frac{\alpha}{\beta}, \quad \text{and} \quad A = \frac{1}{\beta}.$$

The following result about the solutions of Eq.(4.1.6) is extracted from Section 3.4.

Theorem 4.1.1 *Assume that (4.1.2) holds. Then the following statements are true:*

(a) *The positive equilibrium* $\bar{x} = (\alpha - 1)/\beta$ *of Eq.(4.1.6) is locally asymptotically stable if*

$$\frac{\alpha - 1}{\alpha} < 2 \cos \frac{k\pi}{2k + 1}$$

and unstable if

$$\frac{\alpha - 1}{\alpha} > 2\cos\frac{k\pi}{2k+1}.$$

(b) *The positive equilibrium* $\bar{x} = (\alpha - 1)/\beta$ *of Eq.(4.1.6) is globally asymptotically stable if*

$$(\alpha - 1)(k - 1) \leq 1.$$

In particular, the positive equilibrium $\bar{x}$ *of Eq.(4.1.6) is globally asymptotically stable if* $k = 0$ *or* $k = 1$.

(c) *Eq.(4.1.6) is permanent.*

(d) *Every positive solution of Eq.(4.1.6) oscillates about the positive equilibrium* $\bar{x} = (\alpha - 1)/\beta$ *if and only if*

$$\frac{\alpha - 1}{\alpha} > \frac{k^k}{(k+1)^{k+1}}.$$

The study of Eq.(4.1.6) is far from being complete. In particular, it is highly desirable to improve part (b) of Theorem 4.1.1 when $k \geq 2$.

On the basis of computer observations we believe that the following conjecture is true:

Research Project 4.1.1 (Conjecture). *Assume that (4.1.2) holds. Show that the positive equilibrium* $\bar{x} = (\alpha - 1)/\beta$ *of Eq.(4.1.6) is globally asymptotically stable if and only if it is locally asymptotically stable.*

An interesting extension of Eq.(4.1.6) is the delay difference equation

$$x_{n+1} = \frac{\alpha x_n}{1 + \sum_{i=1}^{m} \beta_i x_{n-k_i}}, \quad n = 0, 1, \dots \qquad (4.1.7)$$

where

$$\alpha \in (1, \infty), \quad \beta_1, \dots, \beta_m \in (0, \infty), \quad \text{and} \quad k_1, \dots k_m \in \{0, 1, \dots\}. \qquad (4.1.8)$$

Let $k = \max\{k_1, \dots k_m\}$. If $a_{-k}, \dots, a_0$ are $(k+1)$ given constants such that,

$$a_n \geq 0 \quad \text{for} \quad n = -k, \dots, -1 \quad \text{and} \quad a_0 > 0 \qquad (4.1.9)$$

then Eq.(4.1.7) has a unique positive solution satisfying the initial conditions

$$x_n = a_n \quad \text{for} \quad n = -k, \dots, 0. \qquad (4.1.10)$$

The next theorem gives necessary and sufficient conditions for all positive solutions of Eq.(4.1.8) to oscillate about its unique positive equilibrium

$$\bar{x} = \frac{\alpha - 1}{\displaystyle\sum_{i=1}^{m} \beta_i}.$$

Theorem 4.1.2 *Assume that (4.1.8) holds. Then every solution of Eq.(4.1.7) oscillates about its positive equilibrium $\bar{x}$ if and only if every solution of the linear difference equation*

$$y_{n+1} - y_n + \frac{(\alpha - 1)\displaystyle\sum_{i=1}^{m} \beta_i y_{n-k_i}}{\alpha \displaystyle\sum_{i=1}^{m} \beta_i} = 0, \ n = 0, 1, \ldots \qquad (4.1.11)$$

oscillates about zero.

Proof. The change of variables

$$x_n = \frac{\alpha - 1}{\displaystyle\sum_{i=1}^{m} \beta_i e^{z_n}}, \quad \text{for} \ \ n = 0, 1, \ldots$$

transforms Eq.(4.1.7) to the difference equation

$$z_{n+1} - z_n + f(z_{n-k_1}, \ldots, z_{n-k_m}) = 0, \ n = 0, 1, \ldots \qquad (4.1.12)$$

where

$$f(u_1, \ldots, u_m) = \ln \left(\frac{1}{\alpha} + \frac{\alpha - 1}{\alpha} \frac{\displaystyle\sum_{i=1}^{m} \beta_i e^{u_i}}{\displaystyle\sum_{i=1}^{m} \beta_i} \right).$$

Consider the function $\varphi(x) = -\ln x$. Since φ is a convex function in the interval $(0, \infty)$, it satisfies the well-known Jensen's inequality for convex functions

$$\varphi\left(\sum_{i=0}^{m} q_i t_i \right) \leq \sum_{i=0}^{m} q_i \varphi(t_i)$$

where $q_0, \ldots, q_m \in (0, \infty)$ with $\sum_{i=0}^{m} q_i = 1$ and $t_0, \ldots, t_m \in (0, \infty)$ are arbitrary.

By taking
$$t_0 = 1, \quad \text{and} \quad t_i = e^{u_i} \quad \text{for} \quad i = 1, \ldots, m$$

and
$$q_0 = 1/\alpha \quad \text{and} \quad q_i = \frac{\alpha - 1}{\alpha} \frac{\beta_i}{\sum\limits_{j=1}^{m} \beta_j}, \quad \text{for} \quad i = 1, \ldots, m$$

we obtain
$$\varphi\left(\frac{1}{\alpha} + \frac{\alpha - 1}{\alpha} \frac{\sum\limits_{i=1}^{m} \beta_i e^{u_i}}{\sum\limits_{i=1}^{m} \beta_i}\right) \leq \frac{\varphi(1)}{\alpha} + \frac{\alpha - 1}{\alpha} \frac{\sum\limits_{i=1}^{m} \beta_i \varphi(e^{u_i})}{\sum\limits_{i=1}^{m} \beta_i}.$$

Since $\varphi(1) = 0$, $\varphi(e^{u_i}) = -u_i$ and
$$\varphi\left(\frac{1}{\alpha} + \frac{\alpha - 1}{\alpha} \frac{\sum\limits_{i=1}^{m} \beta_i e^{u_i}}{\sum\limits_{i=1}^{m} \beta_i}\right) = -f(u_1, \ldots, u_m)$$

we have
$$f(u_1, \ldots, u_m) \geq \frac{(\alpha - 1) \sum\limits_{i=1}^{m} \beta_i u_i}{\alpha \sum\limits_{i=1}^{m} \beta_i}.$$

One can now see that all the hypotheses of Theorem 1.2.4 are satisfied, and that the linear equation associated with Eq.(4.1.12) is Eq.(4.1.11). The proof of the theorem is therefore a consequence of Theorem 1.2.4. □

The local asymptotic stability of Eq.(4.1.7) is determined by the linearized equation (4.1.11). For example it follows from Remark 2.3.2 that a sufficient condition for the local stability of the equilibrium $\bar{x}$ of Eq.(4.1.7) is
$$(\alpha - 1) \sum\limits_{i=1}^{m} \beta_i (k_i - 1) < \sum\limits_{i=1}^{m} \beta_i.$$

Also by applying Theorem 3.1.1(b), it follows that every positive solution of Eq.(4.1.7) is bounded away from zero and infinity by positive constants.

Also, from Theorem 3.1.1(c) we have the following condition for global attractivity of $\bar{x}$:
$$(\alpha - 1)(k - 1) \leq 1.$$

Research Project 4.1.2 (Conjecture). *Assume that (4.1.8) holds. Show that the positive equilibrium $\bar{x}$ of Eq.(4.1.7) is globally asymptotically stable if and only if Eq.(4.1.11) is locally asymptotically stable.*

Research Project 4.1.3 *Assume that the sequences $\{\alpha_n\}$ and $\{\beta_n\}$ are periodic. Investigate the behavior of solutions of the rational recursive sequence*

$$x_{n+1} = \frac{\alpha_n x_n}{1 + \beta_n x_{n-1}}, \quad n = 0, 1, \ldots.$$

4.2 A Simple Genotype Selection Model

Our aim in this section is to study the oscillation and the global asymptotic stability of the simple genotype selection model

$$y_{n+1} = \frac{y_n e^{\beta(1-2y_{n-k})}}{1 - y_n + y_n e^{\beta(1-2y_{n-k})}}, \quad n = 0, 1, \ldots \tag{4.2.1}$$

where

$$\beta \in (0, \infty) \quad \text{and} \quad k \in \{0, 1, 2, \ldots\} \tag{4.2.2}$$

and where $y_{-k}, \ldots, y_0 \in [0, 1]$ are arbitrary initial conditions.

When $k = 0$, Eq.(4.2.1) was introduced by May [5] as an example of a map generated by a simple model for frequency dependent natural selection.

The local stability of the positive equilibrium point $y_e = 1/2$ of Eq.(4.2.1), when $k = 0$, was investigated by May [5]. According to May [5, p.540] "for frequency dependent selective forces so strong that $\beta > 4$, this equilibrium point gives way to a stable 2-point cycle, ... and the 2-cycle never becomes unstable". In the next section we will study the existence and stability of periodic orbits of Eq.(4.2.1) when $k = 0$.

The appearance of y_{n-k} in the selection coefficient reflects the fact that the environment at the present time depends upon the activity of the population at some time in the past and that this in turn depends upon the gene frequency at that time. Since different genotypes act differently on the environment, the past genetic makeup can affect possible nesting sites, soil fertility, food supply, predators, etc.

The points 0, 1/2, and 1 are the only equilibrium solutions of Eq.(4.2.1). One can easily see that $y_n \in [0, 1]$ for all $n = 0, 1, \ldots$. Also if $y_N = 0$ for some $N \in \mathbf{N}$, then $y_n = 0$ for all $n \geq 0$ and if $y_N = 1$ for some $N \in \mathbf{N}$, then $y_n = 1$ for all $n \geq 0$. It follows by linearized stability that the points 0 and 1 are unstable equilibria of Eq.(4.2.1).

In the sequel we will restrict our attention to the difference equation

$$y_{n+1} = \frac{y_n e^{\beta(1-2y_{n-k})}}{1 - y_n + y_n e^{\beta(1-2y_{n-k})}}, \ n = 0, 1, \ldots \\ y_{-k}, y_{-k+1}, \ldots, y_0 \in (0, 1). \quad \right\} \tag{4.2.3}$$

Then clearly, $y_n \in (0, 1)$ for all $n \geq -k$. By introducing the substitution

$$x_n = \frac{y_n}{1 - y_n},$$

Eq.(4.2.3) becomes

$$x_{n+1} = x_n \exp(\beta \frac{1 - x_{n-k}}{1 + x_{n-k}}), \ n = 0, 1, \ldots \\ x_{-k}, x_{-k+1}, \ldots, x_0 \in (0, \infty). \quad \right\} \tag{4.2.4}$$

Also, after the substitution $x_n = \exp(z_n)$, Eq.(4.2.4) becomes

$$z_{n+1} - z_n + \beta \tanh(z_{n-k}/2) = 0, \ n = 0, 1, \ldots \\ z_{-k}, z_{-k+1}, \ldots, z_0 \in (-\infty, \infty). \quad \right\} \tag{4.2.5}$$

The following theorem gives explicit necessary and sufficient conditions to insure that every solution of the difference equation (4.2.3) oscillates about $1/2$.

Theorem 4.2.1 *Assume β is a positive real number and k is a nonnegative integer. Then the following statements are true:*

(a) If $k = 0$, then every solution of Eq.(4.2.3) oscillates about $\bar{y} = 1/2$ if and only if $\beta > 2$.

(b) If $k \geq 1$, then every solution of Eq.(4.2.3) oscillates about $\bar{y} = 1/2$ if and only if

$$\beta > 2\frac{k^k}{(k+1)^{k+1}}.$$

Proof. Since the dynamics of Eq.(4.2.3) are equivalent to those of (4.2.5) with $1/2$ replaced by 0, we will consider Eq.(4.2.5). First consider the case where $k + \beta/2 \neq 1$. Observe that the function $f(u) = 2\tanh(u/2)$ is continuous from $\mathbf{R}$ to $\mathbf{R}$, satisfies

$$uf(u) > 0 \ \text{ for } \ u \neq 0, \quad \lim_{u \to 0} \frac{f(u)}{u} = 1,$$

and

$$f(u) \le u \quad \text{for} \quad u \ge 0.$$

Therefore by Theorem 1.2.3, every solution of Eq.(4.2.5) oscillates if and only if every solution of the associated linearized equation

$$w_{n+1} - w_n + \frac{\beta}{2} w_n = 0 \tag{4.2.6}$$

oscillates. By Corollary 1.2.1, Eq.(4.2.6) oscillates if and only if

$$\text{either} \quad k = 0 \quad \text{and} \quad \beta > 2$$

$$\text{or} \qquad k \ge 1 \quad \text{and} \quad \beta > 2 \frac{k^k}{(k+1)^{k+1}}.$$

Finally when $k + \beta/2 = 1$, which is equivalent to $k = 0$ and $\beta = 2$, the result follows from Theorem 1.5.2.   □

Next we study the stability behavior of Eq.(4.2.3) when $k = 0$.

Theorem 4.2.2 *Let $\beta > 0$ and consider the difference equation*

$$\left. \begin{array}{l} y_{n+1} = \dfrac{y_n e^{\beta(1-2y_n)}}{1 - y_n + y_n e^{\beta(1-2y_n)}}, \quad n = 0, 1, \ldots \\[3mm] y_0 \in (0,1). \end{array} \right\} \tag{4.2.7}$$

Then the following statements are true.

(a) If $0 < \beta \le 4$, the equilibrium solution $y_n = 1/2$ is globally asymptotically stable.

(b) If $\beta > 4$, the equilibrium solution $y_n = 1/2$ is unstable.

Proof. The proof will be established in a series of three Lemmas.   □

Lemma 4.2.1 *If $0 < \beta < 4$, then $y_n = 1/2$ is an asymptotically stable solution of Eq.(4.2.7), while if $\beta > 4$, then $y_n = 1/2$ is an unstable solution of Eq.(4.2.7).*

Proof. Set

$$g(x) = \frac{x e^{\beta(1-2x)}}{1 - x + x e^{\beta(1-2x)}}, \quad \text{for} \quad 0 < x < 1. \tag{4.2.8}$$

Then $g'(1/2) = 1 - \beta/2$ and so the result follows by linearized stability.   □

Lemma 4.2.2 *Let* $0 < \beta \le 2$. *Then* $y_n = 1/2$ *is a global attractor of Eq.(4.2.7)*.

Proof. We wish to apply Theorem 1.5.1(a). Consider $g : [0,1) \to [0,1)$ where g is defined by (4.2.8). Then g is continuous, $g(0) = 0$, $g(x) > 0$ if $0 < x < 1$, and g has exactly one positive fixed point $\bar{x} = 1/2$. Note that $g(1/4) > 1/4$ and so $g(x) > x$ if $0 < x < 1/2$. Similarly, $g(3/4) < 3/4$ from which it follows that $g(x) < x$ if $1/2 < x < 1$. Finally, $g'(x_m) = 0$ if and only if $x_m = \frac{1}{2}(1 \pm (1 - \frac{2}{\beta})^{1/2})$. But $1 - 2/\beta > 0$ is equivalent to $\beta > 2$ and we have assumed that $0 < \beta \le 2$. The result follows by part (a) of Theorem 1.5.1. $\qquad\square$

Lemma 4.2.3 *Let* $2 < \beta \le 4$. *Then* $y_n = 1/2$ *is a globally asymptotically stable solution of Eq.(4.2.7)*.

Proof. We shall use Theorem 1.5.3. Consider $g : (0,1) \to (0,1)$ where g is defined by Eq.(4.2.8), and let $V : (0,1) \to [0,\infty)$ be given by $V(x) = (x - 1/2)^2$. Clearly V is continuous, decreasing on $(0, 1/2)$ and increasing on $(1/2, 1)$. If $0 < x < 1$,

$$
\begin{aligned}
\Delta V(x) &= V(g(x)) - V(x) = \left(\frac{xe^{\beta(1-2x)}}{1 - x + xe^{\beta(1-2x)}} - \frac{1}{2}\right)^2 - \left(x - \frac{1}{2}\right)^2 \\
&= -xe^{\beta(1-2x)}\left[\frac{e^{\beta(1-2x)}}{1 - x + xe^{\beta(1-2x)}} - 1\right] \\
&\quad \times \left[\frac{-x}{1 - x + xe^{\beta(1-2x)}} + \frac{(1-x)}{e^{\beta(1-2x)}}\right].
\end{aligned}
$$

Observe that

$$
\frac{e^{\beta(1-2x)}}{1 - x + xe^{\beta(1-2x)}} - 1 = \frac{(1-x)(e^{\beta(1-2x)} - 1)}{1 - x + xe^{\beta(1-2x)}}
\begin{cases}
> 0 & \text{if } 0 < x < \frac{1}{2} \\
= 0 & \text{if } x = \frac{1}{2} \\
< 0 & \text{if } \frac{1}{2} < x < 1.
\end{cases}
$$

Also

$$
\frac{-x}{1 - x + xe^{\beta(1-2x)}} + \frac{(1-x)}{e^{\beta(1-2x)}} = \frac{x(1 - x + xe^{\beta(1/2-x)})(\frac{1-x}{x} - e^{\beta(1/2-x)})}{(1 - x + xe^{\beta(1-2x)})e^{\beta(1-2x)}}.
$$

Since $0 < x < 1$, we need only to determine the sign of $G(x) = \frac{1-x}{x} - e^{\beta(1/2-x)}$.

We shall first show that $G(x) > 0$ if $0 < x < 1/2$. Note that $G(1/2) = 0$, and so it suffices to show that $G'(x) < 0$ for $0 < x < 1/2$. Now

$$
G'(x) = -\frac{1}{x^2}e^{-\beta x}(e^{\beta x} - x^2\beta e^{-\beta/2}).
$$

Set $H(x) = e^{\beta x} - x^2 \beta e^{\beta/2}$. Then $H(0) = 1$ and $H(1/2) = e^{\beta/2}(1 - \beta/4) \geq 0$. It suffices to show that if $0 < x < 1/2$ and $H'(x) = 0$, then $H(x) > 0$. Note that $H'(x) = \beta e^{\beta x} - 2x\beta e^{\beta/2}$. Suppose $H'(x) = 0$, where $0 < x < 1/2$. Then

$$0 = H'(x) = \beta e^{\beta x} - 2x\beta e^{\beta/2} = \beta(e^{\beta x} - 2x e^{\beta/2})$$

and so $e^{\beta x} = 2x e^{\beta/2}$. Hence $H(x) = x e^{\beta/2}(2 - x\beta) > 0$ since $0 < x < 1/2$ and $\beta \leq 4$.

We shall next show that $G(x) < 0$ if $1/2 < x < 1$. Because $G(1/2) = 0$, it suffices to show that $G'(x) < 0$ for $1/2 < x < 1$. But $G'(x) = -H(x)/x^2$ and so we need only show that $H(x) > 0$ for $1/2 < x < 1$. Now $H(1/2) = e^{\beta/2}(1 - \beta/4) \geq 0$, and so it suffices to show $H'(x) > 0$ for $1/2 < x < 1$. But $H'(x) = \beta e^{\beta x} - 2x\beta e^{\beta/2}$, and in particular, $H'(1/2) = 0$. This means that the claim will be established if we can show that

$$H''(x) > 0 \quad \text{for} \quad 1/2 < x < 1.$$

Note that since $\beta > 2$, $H''(x) = \beta^2 e^{\beta x} - 2\beta e^{\beta/2} > 0$ and so $G(x)$ is negative for $1/2 < x < 1$.

Hence $\Delta V(x) < 0$ for all $x \in (0, 1)$ with $x \neq 1/2$, and so we see by Theorem 1.5.3 that $y_n = 1/2$ is a global attractor of Eq.(4.2.7). Thus by Lemma 4.2.1 we see that if $0 < \beta < 4$, then $y_n = 1/2$ is a globally asymptotically stable equilibrium solution of Eq.(4.2.7).

Finally if $\beta = 4$ and $x_0 \in (0, 1)$ with $x_0 \neq 1/2$, then the fact that $\Delta V(x) < 0$ for $x \in (0, 1)$ with $x \neq 1/2$ implies

$$|x_0 - 1/2| > |x_1 - 1/2| > \cdots$$

and so $y_n = 1/2$ is a globally asymptotically stable equilibrium solution in the case $\beta = 4$ also.   □

Remark 4.2.1 It is interesting to note that Theorem 1.5.1 applies only in the case $0 < \beta \leq 2$ while Theorem 1.5.3 applies only in the case $2 < \beta \leq 4$.

Next, consider the delay difference equation (4.2.3) where β is a positive real number and k is a positive integer. We shall first consider the local stability of Eq.(4.2.3).

Lemma 4.2.4 *The equilibrium solution $\bar{y} = 1/2$ of Eq.(4.2.3) is locally asymptotically stable if $0 < \beta < 4\cos\dfrac{k\pi}{2k+1}$ and is unstable if $\beta > 4\cos\dfrac{k\pi}{2k+1}$.*

Proof. The proof follows by linearized stability and Theorem 1.3.6   □

Theorem 4.2.3 *Consider the delay difference equation (4.2.3) where β is a positive real number and k is a positive integer. Then the following statements are true.*

(a) Suppose $k = 1$. Then the equilibrium solution $\bar{y} = 1/2$ is globally asymptotically stable if $0 < \beta \leq 2$ and is unstable if $\beta > 2$.

(b) Suppose $k \geq 2$. Then the equilibrium solution $\bar{y} = 1/2$ is globally asymptotically stable if $0 < \beta \leq 2/k$.

Proof. We will employ Theorem 2.1.1 and Corollary 2.3.1. It is more convenient to study the stability nature of the positive equilibrium solution $\bar{x} = 1$ of the equivalent difference equation (4.2.4). For $x \geq 0$, define $f(x)$ by the formula

$$f(x) = \exp(\beta \frac{1 - x}{1 + x}).$$

When $k = 1$ and $0 < \beta \leq 2$, it follows by Corollary 2.1.1 that $\bar{x} = 1$ is a globally asymptotically stable equilibrium of Eq.(4.2.4). Also linearized stability shows that when $k = 1$ and $\beta > 2$, $\bar{x} = 1$ is an unstable equilibrium of Eq.(4.2.4). The proof of (a) is complete.

Now suppose $k \geq 2$. The hypotheses (a), (b) and (c) of Corollary 2.3.1 are clearly satisfied. Set

$$F(x) = (f(x))^k = \exp\left(k\beta \frac{1 - x}{1 + x}\right).$$

It suffices to verify that the equation $F(F(x)) = x$ has the unique nonnegative solution $\bar{x} = 1$. To this end, consider the difference equation

$$\left.\begin{array}{l} v_{n+1} = F(v_n), n = 0, 1, \ldots \\ \\ v_0 \in [0, \infty). \end{array}\right\} \tag{4.2.9}$$

We shall show that if $0 < \beta \leq 2/k$, then 1 is a global attractor of Eq.(4.2.9) and so the proof that $\bar{y} = 1/2$ is a global attractor will be complete. So assume $0 < \beta \leq 2/k$. Since $F(x)$ is a decreasing function on $[0, \infty)$, it follows that $\bar{v} = 1$ is the unique equilibrium solution of Eq.(4.2.9). Clearly, any solution of Eq.(4.2.9) has the property that $v_n > 0$ for $n \geq 1$. After the substitution $v_n = e^{u_n}$, Eq (4.2.9) is transformed into the equivalent equation

$$\left.\begin{array}{l} u_{n+1} = k\beta \frac{1 - e^{u_n}}{1 + e^{u_n}}, n = 0, 1, \ldots \\ \\ u_0 \in \mathbf{R}. \end{array}\right\} \tag{4.2.10}$$

What we shall actually prove is that the equilibrium solution $\bar{u} = 0$ is a global attractor of Eq.(4.2.10). Consider the function

$$G(u) = k\beta \frac{1 - e^u}{1 + e^u} = -k\beta \tanh \frac{u}{2}.$$

Since $\tanh x$ is an increasing function, $\tanh x < x$ if $x > 0$, $\tanh 0 = 0$, $\tanh x > x$ if $x < 0$, and

$$G(G(u)) = k\beta \tanh(\frac{k\beta}{2} \tanh \frac{u}{2}),$$

it follows easily that

$$G(G(u)) < (\frac{k\beta}{2})^2 u \quad \text{if} \quad u > 0,$$
$$G(G(0)) = 0,$$
$$G(G(u)) > (\frac{k\beta}{2})^2 u \quad \text{if} \quad u < 0.$$

Thus for Eq.(4.2.10) we have that one of the following statements holds: a) If $u_0 = 0$, then $u_n = 0$ for all $n = 0, 1, \ldots$. b) If $u_0 > 0$, then for each $m = 0, 1, \ldots$ we have $u_{2m} > 0$, $u_{2m+1} < 0$,

$$u_{2m+2} < (\frac{k\beta}{2})^2 u_{2m}, \quad \text{and} \quad u_{2m+3} > (\frac{k\beta}{2})^2 u_{2m+1}.$$

c) If $u_0 < 0$, then for each $m = 0, 1, \ldots$ we have $u_{2m} < 0$, $u_{2m+1} > 0$,

$$u_{2m+2} > (\frac{k\beta}{2})^2 u_{2m}, \quad \text{and} \quad u_{2m+3} < (\frac{k\beta}{2})^2 u_{2m+1}.$$

So since $0 < \beta \leq 2/k$, we see that $u_0, u_2, u_4, \ldots$ and $u_1, u_3, u_5, \ldots$ are bounded, monotonically convergent sequences and that

$$\lim_{m \to \infty} u_{2m} = 0 = \lim_{m \to \infty} u_{2m+1}.$$

Thus $\bar{u} = 0$ is a global attractor of Eq.(4.2.10) and so $\bar{x} = 1$ is a global attractor of Eq.(4.2.4); that is, $\bar{y} = 1/2$ is a global attractor of Eq.(4.2.3). Finally, in order to apply Lemma 4.2.4 to complete the proof of statement (b), we must show

$$\frac{2}{k} < 4 \cos \frac{k\pi}{2k + 1} \tag{4.2.11}$$

for every integer $k \geq 2$. With this in mind, consider the inequality

$$\frac{2}{x} < 4 \cos \frac{x\pi}{2x + 1} \quad \text{for} \quad x > 1. \tag{4.2.12}$$

The change of variables $t = \pi/(2x + 1)$ transforms (4.2.12) into the equivalent in-equality

$$\frac{2t}{\pi - 2t} < \sin t, \ 0 < t < \pi/6. \tag{4.2.13}$$

Now $2t/(\pi - 2t)$ is a strictly convex function on $(0, \pi/6)$, and $\sin t$ is a strictly concave function on $(0, \pi/6)$. The proof of inequality (4.2.13) (and hence of (4.2.11)) follows from the observation that

$$\frac{2t}{\pi - 2t} = \sin t \ \text{ for } \ t = \pi/6.$$

The proof is complete. □

In view of Theorem 4.2.3, the condition $0 < \beta \leq 2$ for global asymptotic stability when $k = 1$ is "sharp". Our condition for global asymptotic stability (or even global attractivity) is probably far from sharp when $k \geq 2$. Therefore it is highly desirable to improve part (b) of Theorem 4.2.3 when $k \geq 2$.

On the basis of computer observations we believe that the following conjecture is true:

Research Project 4.2.1 (Conjecture). *Consider the difference equation (4.2.3) where β is positive number and $k \geq 2$. Show that the equilibrium solution $\bar{y} = 1/2$ is globally asymptotically stable if and only if*

$$0 < \beta \leq 4 \cos \frac{k\pi}{2k + 1}.$$

4.3 Periodicity in a Simple Genotype Selection Model

Our aim in this section is to investigate the existence and stability of periodic orbits of the simple genotype selection model (with delay zero),

$$y_{n+1} = \frac{y_n e^{\beta(1-2y_n)}}{1 - y_n + y_n e^{\beta(1-2y_n)}}, \ n = 0, 1, \ldots \tag{4.3.1}$$

where

$$\beta \in (0, \infty) \text{ and } y_0 \in (0, 1). \tag{4.3.2}$$

Clearly $1/2$ is the only equilibrium solution of Eq.(4.3.1) in the interval $(0, 1)$.

Equation (4.3.1) was introduced by May [5] as an example of a map generated by a simple model for frequency dependent natural selection. Eq.(4.3.1) gives the change in gene frequency between the nth generation and the next when the fitness function is $\exp(\beta(1-2y))$.

The local stability of the positive equilibrium point $y_e = 1/2$ of Eq.(4.3.1) was investigated by May [5]. According to May [5, p.540], "for frequency dependent selective forces so strong that $\beta > 4$, this equilibrium point gives way to a stable 2-point cycle, ... and the 2-cycle never becomes unstable".

If $y_0 \in (0,1)$, then $y_n \in (0,1)$ for $n \geq 1$, and the change of variables

$$z_n = \frac{y_n}{1 - y_n}$$

reduces Eq.(4.3.1) to

$$z_{n+1} = z_n \exp(\beta \frac{1 - z_n}{1 + z_n}), \ n = 0, 1, \ldots \tag{4.3.3}$$

with $z_0 \in (0, \infty)$.

Furthermore after the substitution $z_n = e^{x_n}$, Eq.(4.3.3) becomes

$$x_{n+1} - x_n + \beta \tanh(x_n/2) = 0, \ n = 0, 1, \ldots \tag{4.3.4}$$

with $x_0 \in (-\infty, \infty)$. The unique positive equilibrium solution of Eq.(4.3.3) is $z_e = 1$, and the unique equilibrium of Eq.(4.3.4) is $x_e = 0$. These equilibria correspond to the equilibrium solution $y_e = 1/2$ of Eq.(4.3.1). Clearly equations (4.3.1), (4.3.3) and (4.3.4) have equivalent dynamics.

Let h be the function defined by

$$h(y) = \frac{y e^{\beta(1-2y)}}{1 - y + y e^{\beta(1-2y)}} \quad \text{for} \ y \in (0,1).$$

The study of Eq.(4.3.1) is essentially equivalent to the study of the iterates of the function h.

The following concepts related to the iterates of h will be used in the sequel.

For each point $y \in (0,1)$, the **orbit** of y is the sequence of iterates $\{h^n(y)\}$, where $h^0(y) = y$ and $h^{n+1}(y) = h(h^n(h))$. The ω-**limit set of** y is the set of limit points of the orbit of y.

A point p is said to be **periodic of prime period** m if $h^m(p) = p$, while $h^i(p) \neq p$ for $0 < i < m$.

If $|dh^m/dx^m(p)| < 1$, then it can be shown that the point p is **stable**.

Let p be a periodic point of prime period m. The **stable manifold** of p, $S(p)$, is the set

$$S(p) = \{x : \lim_{n \to \infty} h^{mn}(x) = p\}.$$

Let us first summarize some important properties of the function

$$f(u) = u - \beta \tanh(u/2) \tag{4.3.5}$$

and its second iterate

$$F(u) = f(f(u)) = u - \beta \tanh(u/2) - \beta \tanh(u - \beta \tanh(u/2)) \tag{4.3.6}$$

when $\beta > 4$.

Proposition 4.3.1 *Summary of properties of f*

(a) f is an odd function;

(b) Let $\bar{x}_0$ be the unique positive root of the equation

$$x = \beta \tanh(x/2).$$

Then the only zeros of f are 0, $-\bar{x}_0$ and $\bar{x}_0$;

(c) $f(x) > 0$ for $x \in (-\bar{x}_0, 0) \cup (\bar{x}_0, \infty)$;

(d) $f(x) < 0$ for $x \in (-\infty, -\bar{x}_0) \cup (0, \bar{x}_0))$;

(e) The only relative minimum of f is

$$(\bar{x}_m, f(\bar{x}_m)),$$

where

$$\bar{x}_m = \ln((\beta - 1) + \sqrt{\beta^2 - 2\beta}) > 0;$$

(f) The only relative maximum of f is

$$(-\bar{x}_m, -f(\bar{x}_m));$$

(g) $-\beta < -\bar{x}_0 < -\bar{x}_m < 0 < \bar{x}_m < \bar{x}_0 < \beta$;

(h) f is increasing in $(-\infty, -\bar{x}_m) \cup (\bar{x}_m, \infty)$;

(i) f is decreasing in $(-\bar{x}_m, \bar{x}_m)$;

(j) $f(x) > x - \bar{x}_0$ for $0 \le x \le \bar{x}_m$;

(k) $0 > f(\bar{x}_m) > \bar{x}_m - \bar{x}_0 > -\bar{x}_0$;

(l) $f([-\bar{x}_0, \bar{x}_0]) = f([-\bar{x}_m, \bar{x}_m]) \subset [-\bar{x}_0, \bar{x}_0]$;

(m) $f(x) > x - \beta$ and $f(x) \sim x - \beta$ as $x \to \infty$;

Proposition 4.3.2   *Summary of properties of F*

(a) F is an odd function;

(b) Let $\bar{x}$ be the unique positive solution of the equation

$$x = \frac{\beta}{2}\tanh(x/2).$$

Then $-\bar{x}$, 0 and $\bar{x}$ are the only fixed points of F;

(c) $f(\bar{x}) = -\bar{x}$;

(d) Let $\bar{x}_1 = f(\bar{x}_0) > 0$. Then the only zeros of F are the following five points: 0, $-\bar{x}_0$, $\bar{x}_0$, $-\bar{x}_1$ and $\bar{x}_1$;

(e) $F(x) > 0$ for $x \in (-\bar{x}_1, -\bar{x}_0) \cup (0, \bar{x}_0) \cup (\bar{x}_1, \infty)$;

(f) $F(x) < 0$ for $x \in (-\infty, -\bar{x}_1) \cup (-\bar{x}_0, 0) \cup (\bar{x}_0, \bar{x}_1)$;

(g) $F(x) > x - 2\beta$ and $F(x) \sim x - 2\beta$ as $x \to \infty$;

(h) $0 \le F'(\bar{x}) = F'(-\bar{x}) = (1 - 2\bar{x}/\sinh \bar{x})^2 < 1$.

Proposition 4.3.3   *Additional properties of F when* $\sinh \bar{x} \le 2\bar{x}$

(a) The equation

$$f(x) = \bar{x}_m > 0.$$

has a unique positive solution $\bar{x}_r$. Moreover, the only relative minima of F are the following two points:

$$(-\bar{x}_m, F(\bar{x}_m)) \quad \text{and} \quad (\bar{x}_r, f(\bar{x}_m));$$

(b) The only relative maxima of F are the following two points:

$$(-\bar{x}_r, -f(\bar{x}_m)) \quad \text{and} \quad (\bar{x}_m, F(\bar{x}_m));$$

(c) F is increasing in $(-\infty, -\bar{x}_r) \cup (-\bar{x}_m, \bar{x}_m) \cup (\bar{x}_r, \infty)$;

(d) F is decreasing in $(-\bar{x}_r, -\bar{x}_m) \cup (\bar{x}_m, \bar{x}_r)$;

(e) $0 < \bar{x} \le \bar{x}_m < \bar{x}_0 < \bar{x}_r < \bar{x}_1$;

(f) $f(\bar{x}_m) \ge -\bar{x}_m$;

(g) $-1 < f'(\bar{x}) \le 0$.

Proposition 4.3.4 *Additional properties of F when* $\sinh \bar{x} > 2\bar{x}$

(a) Let $\bar{x}_p$, $\bar{x}_q$ and $\bar{x}_r$, with $0 < \bar{x}_p < \bar{x}_q < \bar{x}_r$, denote the three positive roots of the equation

$$f(x) = \bar{x}_m > 0.$$

Then the only relative minima of F are the following four points:

$$(-\bar{x}_p, f(\bar{x}_m)), \ (-\bar{x}_q, f(\bar{x}_m)), \ (\bar{x}_m, \ F(\bar{x}_m)), \ \text{and} \ (\bar{x}_r, f(\bar{x}_m));$$

(b) The only relative maxima of F are the following four points:

$$(-\bar{x}_r, -f(\bar{x}_m)), \ (-\bar{x}_m, -F(\bar{x}_m)), \ (\bar{x}_p, -f(\bar{x}_m)), \ \text{and} \ (\bar{x}_q, -f(\bar{x}_m));$$

(c) $0 < \bar{x}_p < \bar{x}_m < \bar{x} < \bar{x}_q < \bar{x}_0 < \bar{x}_r < \bar{x}_1$;

(d) F is increasing in $(-\infty, -\bar{x}_r) \cup (-\bar{x}_q, -\bar{x}_m) \cup (-\bar{x}_p, \bar{x}_p) \cup (\bar{x}_m, \bar{x}_q) \cup (\bar{x}_r, \infty)$;

(e) F is decreasing in $(-\bar{x}_r, -\bar{x}_q) \cup (-\bar{x}_m, -\bar{x}_p) \cup (\bar{x}_p, \bar{x}_m) \cup (\bar{x}_q, \bar{x}_r)$;

(f) $f(\bar{x}_m) < -\bar{x}_m$;

(g) $0 < f'(\bar{x}) < 1$.

Remark 4.3.1 As we see, the function F has different behavior for different values of β. The approximate value of β_0 for which we have $2\bar{x} = \sinh \bar{x}$ is (by Newton's method) $\beta_0 = 5.46783\ldots$.

We will now study the properties of the periodic orbits of the difference equation

$$x_{n+1} - x_n + \beta \tanh(x_n/2) = 0, \ n = 0, 1, \ldots \tag{4.3.7}$$

where $\beta > 4$ and where the initial condition x_0 is an arbitrary real number.

Lemma 4.3.1 *Assume that* $\beta > 4$ *and let* $\bar{x}$ *denote the unique positive solution of the equation*

$$x = \frac{\beta}{2} \tanh(x/2). \tag{4.3.8}$$

Then $\{\bar{x}, -\bar{x}, \bar{x}, -\bar{x}, \ldots\}$ *and* $\{-\bar{x}, \bar{x}, -\bar{x}, \bar{x}, \ldots\}$ *are periodic solutions of Eq. (4.3.7), each with period two.*

Proof. The existence of the periodic solutions follows from Proposition 2.2(b). □

Lemma 4.3.2 *Consider Eq.(4.3.7) with $\beta > 4$. Let $\bar{x}_0$ be as defined in Proposition 2.2(b). Then the following statements are true:*

(a) For every $x_0 \in \mathbf{R}$, there exists a positive integer n_0 such that

$$x_n \in [-\bar{x}_0, \bar{x}_0] \quad \text{for} \quad n \geq n_0. \tag{4.3.9}$$

(b) Let $P_f = \{x : f^n(x) = 0 \quad \text{for} \quad n = 1, 2, \ldots\}$. Then P_f is a countable set. Furthermore,

$$P_f \cap (-\bar{x}_0, 0) = \emptyset \quad \text{and} \quad P_f \cap (0, \bar{x}_0) = \emptyset. \tag{4.3.10}$$

Proof. (a) When $x_0 \in [-\bar{x}_0, \bar{x}_0]$, the result follows from Proposition 2.1(1). Now consider the case where $x_0 > \bar{x}_0$. The case where $x_0 < -\bar{x}_0$ is similar and will be omitted. Since $f(x) < x$ for $x > 0$, it follows that the sequence $\{x_n\}$ is decreasing. If for some integer n_0, $x_{n_0-1} > \bar{x}_0$ and $x_{n_0} \leq \bar{x}_0$, then we have $x_{n_0} = f(x_{n_0-1}) > f(\bar{x}_m) > -\bar{x}_0$ and (4.3.9) is proved. Assume for the sake of contradiction that $x_n > \bar{x}_0$ for all $n \geq 0$. Then $\{x_n\}$ converges and

$$\lim_{n \to \infty} x_n = x' \geq \bar{x}_0 > 0.$$

Letting $n \to \infty$ in (4.3.7) we obtain $x' = 0$ and this contradiction completes the proof.

(b) Since $f(x) < 0$ for $x \in (0, \bar{x}_0)$ and $f(x) > 0$ for $x \in (-\bar{x}_0, 0)$, it follows that $f^n(x) \neq 0$ for $n = 1, 2, \ldots$ and $x \in (-\bar{x}_0, 0) \cup (0, \bar{x}_0)$ and so (4.3.10) holds.

Next, it is obvious that $-\bar{x}_0, 0, \bar{x}_0 \in P_f$. Consider the sequences $\{\bar{x}_n\}$ and $\{-\bar{x}_n\}$, where $\bar{x}_n$ is the unique positive solution of the equation

$$f(\bar{x}_n) = \bar{x}_{n-1} \quad \text{for} \quad n = 1, 2, \ldots. \tag{4.3.11}$$

Clearly,

$$P_f = \{0\} \cup \{\bar{x}_0, \bar{x}_1, \ldots\} \cup \{-\bar{x}_0, -\bar{x}_1, \ldots\}$$

and P_f is a countable set. Since $f(x) < x$ for $x > \bar{x}_0$, it follows that the sequence $\{\bar{x}_n\}$ is increasing and that

$$\lim_{n \to \infty} \bar{x}_n = \infty.$$

The proof is complete.   □

Lemma 4.3.3 *(a) If $x_0 \in P_f$, then $\{x_n\}$ is eventually zero.*

(b) If $x_0 \in (-\infty, \infty) \setminus P_f$, then there exists a positive integer n_1 such that

$$x_n \in (\bar{x}_m - \bar{x}_0, 0) \cup (0, \bar{x}_0 - \bar{x}_m) \quad \text{for} \quad n \geq n_1. \tag{4.3.12}$$

(c) Let $x_0 \in (-\infty, \infty) \setminus P_f$. Then there exists a positive integer n_2 such that for $n \geq n_2$,

$$\text{either} \quad x_{2n} \in (0, \bar{x}_0 - \bar{x}_m) \quad \text{and} \quad x_{2n+1} \in (\bar{x}_m - \bar{x}_0, 0) \quad (4.3.13)$$

$$\text{or} \quad x_{2n} \in (\bar{x}_m - \bar{x}_0, 0) \quad \text{and} \quad x_{2n+1} \in (0, \bar{x}_0 - \bar{x}_m). \quad (4.3.14)$$

Proof. Part (a) is a direct consequence of Lemma 4.3.2(b).

(b) From Lemma 4.3.3(a) and (b) it follows that if $x_0 \in (-\infty, \infty) \setminus P_f$, then there exists a positive integer n_0 such that $x_{n_0} \in (-\bar{x}_0, \bar{x}_0)$.

First consider the case where $x_{n_0} > 0$. Then by using Proposition 2.1(k) we have

$$0 > x_{n_0+1} = f(x_{n_0}) > f(x_m) > \bar{x}_m - \bar{x}_0 > -\bar{x}_0.$$

Similarly if $x_{n_0} < 0$, then

$$0 < x_{n_0+1} = f(x_{n_0}) < -f(x_m) < \bar{x}_0 - \bar{x}_m < \bar{x}_0$$

and (4.3.12) holds for $n_1 = n_0 + 1$.

(c) From (4.3.12) it follows that for some positive integer n_1,

$$\text{either} \quad x_{n_1} \in (\bar{x}_m - \bar{x}_0, 0) \quad \text{or} \quad x_{n_1} \in (0, \bar{x}_0 - \bar{x}_m).$$

We will only consider the case where $x_{n_1} \in (0, \bar{x}_0 - \bar{x}_m)$. The other case is similar and will be omitted. Then $x_{n_1+1} = f(x_{n_1}) \in (\bar{x}_m - \bar{x}_0, 0)$. Similarly we have $x_{n_1+2} = f(x_{n_1+1}) \in (0, \bar{x}_0 - \bar{x}_m)$ which implies that

$$x_{n_1+2k} \in (0, \bar{x}_0 - \bar{x}_m) \quad \text{and} \quad x_{n_1+2k+1} \in (\bar{x}_m - \bar{x}_0, 0) \quad \text{for} \quad k = 0, 1, \ldots.$$

The proof is complete. $\qquad\square$

The following technical lemma will be useful in the sequel.

Lemma 4.3.4 *Consider the difference equation*

$$y_{n+1} = g(y_n), \quad n = 0, 1, \ldots \quad (4.3.15)$$

with $y_0 \in (0, \infty)$ and where g is defined as follows:

$$g(x) = \begin{cases} F(x) & \text{for} \quad 0 \leq x \leq a \\ F(a) & \text{for} \quad x > a \end{cases} \quad (4.3.16)$$

where

$$a = \begin{cases} \bar{x}_0 - \bar{x}_m & \text{for} \quad \sinh \bar{x} \leq 2\bar{x} \\ \max\{(\bar{x}_0 - \bar{x}_m), \bar{x}_q\} & \text{for} \quad \sinh \bar{x} > 2\bar{x}. \end{cases} \quad (4.3.17)$$

Then $\bar{x}$ is a global attractor of all positive solutions of Eq.(4.3.15).

Proof. We wish to apply Theorem 1.5.1. Clearly $g \in \mathbf{C}[[0,\infty),[0,\infty)]$ and $g(0) = F(0) = 0$. Next we will show that g has a unique positive equilibrium $\bar{x}$. Since $\bar{x} \in (0,\bar{x}_0)$, we have

$$\bar{x}_m - \bar{x}_0 < f(\bar{x}) = -\bar{x} < 0$$

and so $\bar{x} \in (0,\bar{x}_0 - \bar{x}_m)$. Since $\bar{x}$ is the only positive fixed point of F, it follows that $\bar{x}$ is also a fixed point of g. Since $F(x) > x$ for $x \in (0,\bar{x})$ and $F(x) < x$ for $x > \bar{x}$, it follows that

$$g(x) > x \quad \text{for} \quad x \in (0,\bar{x}) \quad \text{and} \quad g(x) < x \quad \text{for} \quad x \in (\bar{x},\infty)$$

which shows that g has no positive fixed points other than $\bar{x}$.

First consider the case where $\sinh \bar{x} \leq 2\bar{x}$. Since $\bar{x} \leq \bar{x}_m$, it follows that g is increasing in $(0,\bar{x})$ while $g(x) \leq \bar{x}$ for $x \in [0,\bar{x})$. The global attractivity of $\bar{x}$ follows by using Theorem 1.5.1(b).

Next, consider the case where $\sinh \bar{x} > 2\bar{x}$. We will apply Theorem 1.5.1. Since the function g attains its absolute maximum in $[0,\bar{x}]$ at the point $\bar{x}_p$ and

$$g(\bar{x}_p) = F(\bar{x}_p) = -f(\bar{x}_m) > -f(\bar{x}) = \bar{x},$$

it remains to show that $g(g(x)) > x$ for $x \in [\bar{x}_p, \bar{x}_m]$.

Let $x \in [\bar{x}_p, \bar{x}_m)$. As F is decreasing on $(\bar{x}_p, \bar{x}_m)$ and $F(x) > x$ for $x \in (0,\bar{x})$, we have

$$\bar{x}_m < F(\bar{x}_m) < F(x) \leq F(\bar{x}_p) = -f(\bar{x}_m).$$

Since $F(x) < x$ for $x > \bar{x}$, and $\bar{x} < \bar{x}_q$, we have

$$F(x) < F(\bar{x}_p) = F(\bar{x}_q) < \bar{x}_q.$$

Therefore for every $x \in [\bar{x}_p, \bar{x}_m)$, there exists $x' \in (\bar{x}_m, \bar{x}_q]$, such that

$$F(x) = x'.$$

Then since F is increasing in $[\bar{x}_m, \bar{x}_q]$, we obtain

$$F(F(x)) = F(x') > F(\bar{x}_m) > \bar{x}_m > x.$$

Now let $x \in [\bar{x}_m, \bar{x})$. As F is increasing and $F(x) > x$ we have

$$F([\bar{x}_m, \bar{x})) = [F(\bar{x}_m), \bar{x}) \subset [\bar{x}_m, \bar{x})$$

and so

$$F(F(x)) > F(x) > x.$$

Since the functions g and F, as well as g^2 and F^2, coincide in the interval $[\bar{x}_p, \bar{x})$, it follows that the conditions of Theorem 1.5.1(b) are satisfied and we have

$$\lim_{n \to \infty} y_n = \bar{x}. \tag{4.3.18}$$

The proof is complete.   □

Lemma 4.3.5 *Consider Eq.(4.3.7) with $\beta > 4$, and assume that the initial condition $x_0 \in (-\infty, \infty) \setminus P_f$. Then*

$$\text{either} \quad \lim_{n \to \infty} x_{2n} = \bar{x} \quad \text{and} \quad \lim_{n \to \infty} x_{2n+1} = -\bar{x} \tag{4.3.19}$$

$$\text{or} \quad \lim_{n \to \infty} x_{2n} = -\bar{x} \quad \text{and} \quad \lim_{n \to \infty} x_{2n+1} = \bar{x}. \tag{4.3.20}$$

Proof. From Lemma 4.3.3 it follows that there exists a positive integer n_2 such that for $n \geq n_2$,

$$\text{either} \quad x_{2n} \in (0, \bar{x}_0 - \bar{x}_m) \quad \text{or} \quad x_{2n} \in (\bar{x}_m - \bar{x}_0, 0).$$

We will consider only the case where $x_{2n} \in (0, \bar{x}_0 - \bar{x}_m)$ for $n \geq n_2$. The proof in the other case is similar and will be omitted. Define the sequence $\{y_n\}$ as a solution of Eq.(4.3.15) with initial condition $y_0 = x_{2n_2}$. In view of the properties of the functions f and F and the fact that $x_{2n+2} = F(x_{2n})$, it is easy to see that

$$y_k = x_{2n_2+2k} \quad \text{for} \quad k = 0, 1, \dots.$$

From (4.3.18) it follows that

$$\lim_{n \to \infty} x_{2n} = \bar{x}.$$

Furthermore since $x_{2n+1} = f(x_{2n})$, by using the continuity of f we find

$$\lim_{n \to \infty} x_{2n+1} = f(\bar{x}) = -\bar{x}$$

and the proof is complete. $\qquad \Box$

Lemma 4.3.6 *Consider Eq.(4.3.7) with $\beta > 4$. Let $\{\bar{x}_n\}$ be the sequence defined by (4.3.11). Assume that k is a nonnegative integer. Then the following statements are true:*

(a) If $x_0 \in (\bar{x}_{2k}, \bar{x}_{2k+1})$, then

$$x_{2n} \in \begin{cases} (\bar{x}_{2k-2n}, \bar{x}_{2k-2n+1}) & \text{for} \ \ 0 \leq n \leq k \\ (-\bar{x}_0, 0) & \text{for} \ \ n > k \end{cases} \tag{4.3.21}$$

$$x_{2n+1} \in \begin{cases} (\bar{x}_{2k-2n-1}, \bar{x}_{2k-2n}) & \text{for} \ \ 0 \leq n < k \\ (0, \bar{x}_0) & \text{for} \ \ n \geq k. \end{cases} \tag{4.3.22}$$

(b) If $x_0 \in (\bar{x}_{2k+1}, \bar{x}_{2k+2})$, then

$$x_{2n} \in \begin{cases} (\bar{x}_{2k-2n+1}, \bar{x}_{2k-2n+2}) & \text{for } 0 \le n \le k \\ (0, \bar{x}_0) & \text{for } n > k \end{cases} \tag{4.3.23}$$

$$x_{2n+1} \in \begin{cases} (\bar{x}_{2k-2n}, \bar{x}_{2k-2n+1}) & \text{for } 0 \le n \le k \\ (-\bar{x}_0, 0) & \text{for } n > k. \end{cases} \tag{4.3.24}$$

(c) If $x_0 \in (-\bar{x}_{2k+1}, -\bar{x}_{2k})$, then

$$x_{2n} \in \begin{cases} (-\bar{x}_{2k-2n+1}, -\bar{x}_{2k-2n}) & \text{for } 0 \le n \le k \\ (0, \bar{x}_0) & \text{for } n > k \end{cases} \tag{4.3.25}$$

$$x_{2n+1} \in \begin{cases} (-\bar{x}_{2k-2n}, -\bar{x}_{2k-2n-1}) & \text{for } 0 \le n < k \\ (-\bar{x}_0, 0) & \text{for } n \ge k. \end{cases} \tag{4.3.26}$$

(d) If $x_0 \in (-\bar{x}_{2k+2}, -\bar{x}_{2k+1})$, then

$$x_{2n} \in \begin{cases} (-\bar{x}_{2k-2n+2}, -\bar{x}_{2k-2n+1}) & \text{for } 0 \le n \le k \\ (-\bar{x}_0, 0) & \text{for } n > k \end{cases} \tag{4.3.27}$$

$$x_{2n+1} \in \begin{cases} (-\bar{x}_{2k-2n+1}, -\bar{x}_{2k-2n}) & \text{for } 0 \le n \le k \\ (0, \bar{x}_0) & \text{for } n \ge k. \end{cases} \tag{4.3.28}$$

Proof. We will only prove part (a). The proofs of (b), (c), and (d) are similar and will be omitted.

(a) By using (4.3.11), the increasing nature of $\{\bar{x}_n\}$, and the fact that F is increasing for $x > \bar{x}_0$, it follows that

$$\bar{x}_{2k-2} = F(\bar{x}_{2k}) < x_2 = F(x_0) < F(\bar{x}_{2k+1}) = \bar{x}_{2k-1}.$$

Similarly,

$$x_{2n} \in (\bar{x}_{2k-2n}, \bar{x}_{2k-2n+1}) \quad \text{for } 0 \le n \le k.$$

Since $x_{2k} \in (\bar{x}_0, \bar{x}_1)$ and the function F maps $(\bar{x}_0, \bar{x}_1)$ into $(-\bar{x}_0, 0)$ and $(-\bar{x}_0, 0)$ into itself, it follows that

$$x_{2n} \in (-\bar{x}_0, 0) \quad \text{for } n > k.$$

Next, for $0 \le n < k$ we have

$$\bar{x}_{2k-2n-1} = f(\bar{x}_{2k-2n}) < x_{2n+1} = f(x_{2n}) < f(\bar{x}_{2k-2n+1}) = \bar{x}_{2k-2n}.$$

Also since f maps $(-\bar{x}_0, 0)$ into $(0, \bar{x}_0)$, it follows that

$$x_{2n+1} = f(x_{2n}) \in (0, \bar{x}_0) \quad \text{for} \quad n \geq k$$

and the proof is complete. □

Finally we will study the stability of the periodic orbits of Eq.(4.3.1). Consider the function h which is defined by

$$h(y) = \frac{y e^{\beta(1-2y)}}{1 - y + y e^{\beta(1-2y)}} \quad \text{for} \quad y \in (0,1) \tag{4.3.29}$$

and its second iterate $H(y) = h(h(y))$.

It is easy to see that

$$h(y) = \frac{\exp(f(\ln \frac{y}{1-y}))}{1 + \exp(f(\ln \frac{y}{1-y}))} \quad \text{for} \quad y \in (0,1) \tag{4.3.30}$$

and that

$$f(x) = \ln \left(\frac{h(\frac{e^x}{1+e^x})}{1 - h(\frac{e^x}{1+e^x})} \right) \quad \text{for} \quad x \in (-\infty, \infty) \tag{4.3.31}$$

and so results similar to the results for Eq.(4.3.3) hold for Eq.(4.3.1).

The set P_h corresponding to P_f is given by

$$P_h = \{1/2\} \cup \{1/2 + \bar{y}_0, 1/2 + \bar{y}_1, \ldots\} \cup \{1/2 - \bar{y}_0, 1/2 - \bar{y}_1, \ldots\} \tag{4.3.32}$$

where $\bar{y}_n \in (0, 1/2)$ is the unique solution of the equation

$$h(1/2 + \bar{y}_n) = \begin{cases} 1/2 & \text{for } n = 0 \\ \\ 1/2 + \bar{y}_{n-1} & \text{for } n = 1, 2, \ldots. \end{cases} \tag{4.3.33}$$

Clearly

$$\lim_{n \to \infty} \bar{y}_n = 1/2.$$

Furthermore, the points corresponding to $\bar{x}$ and $-\bar{x}$ are $(1/2 + \bar{y})$ and $(1/2 - \bar{y})$, respectively, where

$$\bar{y} = \frac{e^{\bar{x}} - 1}{2(e^{\bar{x}} + 1)} \in (0, 1/2).$$

Also $\bar{y}$ is the unique solution in the interval $(0, 1/2)$ of the equation

$$\frac{1 + 2\bar{y}}{1 - 2\bar{y}} = e^{\beta \bar{y}}. \tag{4.3.34}$$

The following theorem summarizes the properties of the orbits of the points in $(0, 1)$.

Theorem 4.3.1 *Assume $\beta > 4$ and consider the function h which is defined by (4.3.29). Then the following statements are true:*

(a) $1/2$ is an unstable fixed point of h;

(b) $(1/2 + \bar{y})$ and $(1/2 - \bar{y})$ are stable periodic points of h each of prime period 2;

(c) The orbits of the points in P_h are eventually $1/2$.

(d) The stable manifolds of the points $(1/2 + \bar{y})$ and $(1/2 - \bar{y})$ are, respectively, given by

$$S(1/2 + \bar{y}) \;=\; \bigcup_{k=0}^{\infty}(1/2 - \bar{y}_{2k+1}, 1/2 - \bar{y}_{2k}) \cup (1/2, 1/2 + \bar{y})$$

$$\cup \;\; \bigcup_{k=0}^{\infty}(1/2 + \bar{y}_{2k+1}, 1/2 + \bar{y}_{2k+2})$$

and

$$S(1/2 - \bar{y}) \;=\; \bigcup_{k=0}^{\infty}(1/2 - \bar{y}_{2k+2}, 1/2 - \bar{y}_{2k+1}) \cup (1/2 - \bar{y}, 0)$$

$$\cup \;\; \bigcup_{k=0}^{\infty}(1/2 + \bar{y}_{2k}, 1/2 + \bar{y}_{2k+1}).$$

(e) $(0,1) = P_h \cup S(1/2 + \bar{y}) \cup S(1/2 - \bar{y})$.

Proof. The proof follows directly from Lemmas 4.3.3, 4.3.5 and 4.3.6 and the fact that Eq.(4.3.1) and Eq.(4.3.4) have equivalent dynamics. $\square$

The following theorem describes the behavior of the periodic points $(1/2 + \bar{y})$ and $(1/2 - \bar{y})$ in terms of β.

Theorem 4.3.2 *Assume $\beta > 4$ and let $\bar{y}$ be the unique solution of Eq.(4.3.33) in the interval $(0, 1/2)$. Then the following statements are true.*

(a) $\beta(\bar{y}) = \dfrac{1}{\bar{y}} \ln(\dfrac{1 + 2\bar{y}}{1 - 2\bar{y}})$;

(b) $\bar{y}$ is a continuous and increasing function of β;

(c) $\lim\limits_{\beta \to \infty} \bar{y} = 1/2$;

(d) $\lim\limits_{\beta \to 4+} \bar{y} = 0$.

Proof. (a) This follows directly from (4.3.34). It is easy to see that $\beta'(\bar{y}) > 0$ and so $\beta(\bar{y})$ is an increasing function. Clearly, $\bar{y}$ is also an increasing function of β. Observe that

$$\lim_{\bar{y} \to 1/2-} \beta(\bar{y}) = \infty, \quad \lim_{\bar{y} \to 0+} \beta(\bar{y}) = 4,$$

and that $\bar{y}$ is a continuous and increasing function of β.
The proofs of (b), (c), and (d) are now obvious. □

Research Project 4.3.1 *Study the periodic nature of solutions of the difference equation (4.2.1) when k is a positive integer.*

4.4 A Model of the Spread of an Epidemic

The difference equation

$$\log \frac{1}{x_{n+1}} = \sum_{j=0}^{n} (1 + \epsilon - x_{n-j}) a_j, \quad n = 0, 1, \ldots \tag{4.4.1}$$

describes the spread of an epidemic. See Kelley and Peterson [1, p.87] and Lauwerier [1, Chapter 8].

In Eq.(4.4.1), x_n denotes the fraction of susceptible individuals in the population at the nth day of the epidemic, $a_j > 0$ is a measure of how infectious the sick individuals are at the jth day, and ϵ is a small positive constant.

We will assume that

$$\sum_{n=0}^{\infty} a_n = a < \infty. \tag{4.4.2}$$

The change of variables $x_n = e^{y_n}$ transforms Eq.(4.4.1) to a so-called Volterra difference equation of convolution type

$$y_{n+1} = \sum_{j=0}^{n} a_{n-j}(1 + \epsilon - e^{-y_j}), n = 0, 1, \ldots. \tag{4.4.3}$$

Clearly, $x_n \in [0, 1]$ and so $y_n \geq 0$. It is easy to see that if $y_0 \geq 0$, then $y_n > 0$ for $n = 1, 2, \ldots$.

During the early stages of the epidemic, x_n is near 1, so y_n is near zero and it is reasonable to approximate Eq.(4.4.3) by the linear equation

$$z_{n+1} = \sum_{j=0}^{n} a_{n-j}(\epsilon + z_j), n = 0, 1, \ldots \tag{4.4.4}$$

with

$$z_0 = 0.$$

In some special cases it is possible to find an explicit solution of Eq.(4.4.4).

Before we can establish the first result in this section we need to recall the concept of the z-transform of a sequence. See Györi and Ladas [2] or Kelley and Peterson [1].

Definition 4.4.1 *Let $\{x_n\}$ be a sequence of real numbers defined for $n = 0, 1, \ldots$. The z-transform of this sequence is denoted by $Z(x_n)$ and is defined to be the series*

$$Z(x_n) = \sum_{n=0}^{\infty} \frac{x_n}{z^n}. \tag{4.4.5}$$

The z-transform $Z(x_n)$ is defined for all values of the complex variable z for which the series (4.4.5) converges.

One can show that the following properties are true:

(a) Let $k \in \{1, 2, \ldots\}$. Then

$$Z(x_{n+k}) = z^k Z(x_n) - \sum_{n=0}^{k-1} x_n z^{k-n}.$$

(b) (Convolution property) Let $\{a_n\}$ and $\{b_n\}$ be sequences of real numbers defined for $n = 0, 1, \ldots$. Then

$$Z(\sum_{j=0}^{n} a_j b_{n-j}) = Z(a_n) Z(b_n).$$

Lemma 4.4.1 *Assume that*

$$a_j = ca^j \quad \text{for} \quad j = 0, 1, \ldots \tag{4.4.6}$$

where

$$a, c, a + c \in (0, 1).$$

Let $\{z_n\}$ be the solution of Eq.(4.4.4) with $z_0 = 0$. Then

$$z_n = \frac{\epsilon c}{1 - (a + c)}[1 - (a + c)^n] \quad \text{for} \quad n = 0, 1, \ldots \tag{4.4.7}$$

Proof. By taking z-transforms of both sides of Eq.(4.4.4) we find

$$Z(\sum_{j=0}^{n} ca^j(\epsilon + z_{n-j})) = Z(\epsilon + z_n)Z(ca^n) = [\frac{\epsilon z}{z-1} + Z(z_n)]\frac{cz}{z-a}$$

and so

$$Z(z_n) = \frac{\epsilon c z}{(z-(a+c))(z-1)}$$

from which (4.4.7) follows. $\square$

In particular, this shows that $\{z_n\}$ remains small for all n, so the outbreak does not reach epidemic proportions.

The following comparison results may be useful in studying the properties of solutions of Eq.(4.4.3).

Lemma 4.4.2 *Let $\{y_n\}$ and $\{z_n\}$ be solutions of the equations (4.4.3) and (4.4.4), respectively, such that*

$$0 \le y_0 \le z_0.$$

Then

$$0 \le y_n \le z_n \quad \text{for} \quad n = 0, 1, \dots.$$

Proof. The proof is by induction. Since $y_0 \le z_0$, suppose that $y_k \le z_k$ for $k = 1, 2, \dots n$. Then since $1 + \epsilon - e^{-y} \le \epsilon + y$ for $y \ge 0$, we find

$$\bar{y}_{n+1} = \sum_{j=0}^{n} a_{n-j}(1 + \epsilon - e^{-\bar{y}_j}) \le \sum_{j=0}^{n} a_{n-j}(\epsilon + y_j)$$

$$\le \sum_{j=0}^{n} a_{n-j}(\epsilon + z_j) = z_{n+1}$$

which completes the proof. $\square$

Lemma 4.4.3 *Assume that (4.4.2) holds and let $\{\bar{y}_n\}$ be the solution of Eq.(4.4.3) with initial condition $\bar{y}_n = 0$. Then the following statements are true:*

(a) All positive solutions of Eq.(4.4.3) are bounded. More precisely,

$$y_n \le a(1 + \epsilon) \quad \text{for} \quad n = 1, 2, \dots. \tag{4.4.8}$$

(b) Let $\{\bar{y}_n\}$ be the solution of Eq.(4.4.3) with initial condition $\bar{y}_0 = 0$. Then,

$$\bar{y}_n \leq y_n \quad \text{for} \quad n = 1, 2, \ldots \tag{4.4.9}$$

where $\{y_n\}$ is any solution of Eq.(4.4.3) with $y_0 \geq 0$.

Proof. (a) Since $y_j \geq 0$, from Eq.(4.4.3) it follows that

$$y_{n+1} = \sum_{j=0}^{n} a_{n-j}(1 + \epsilon - e^{-y_j}) \leq \sum_{j=0}^{n} a_{n-j}(1 + \epsilon)$$
$$\leq (1 + \epsilon)a \quad \text{for} \quad n = 0, 1, \ldots$$

which completes the proof of (a).

(b) The proof is by induction. Since $\bar{y}_0 = 0 \leq y_0$, suppose that $\bar{y}_k \leq y_k$ for $k = 1, 2, \ldots n$. Then since the function $1 + \epsilon - e^{-y}$ is increasing, we find

$$\bar{y}_{n+1} = \sum_{j=0}^{n} a_{n-j}(1 + \epsilon - e^{-\bar{y}_j}) \leq \sum_{j=0}^{n} a_{n-j}(1 + \epsilon - e^{-y_j}) = y_{n+1}$$

and the proof is complete. $\qquad\qquad\qquad\qquad\qquad\qquad\qquad\qquad\qquad\square$

Theorem 4.4.1 *Assume that (4.4.2) holds with $a < 1$, and let $\{\bar{y}_n\}$ be the solution of Eq.(4.4.3) with initial condition $\bar{y}_0 = 0$. Then $\{\bar{y}_n\}$ is a globally asymptotically stable solution of Eq.(4.4.3).*

Proof. The substitution $w_n = y_n - \bar{y}_n$ transforms Eq.(4.4.3) into

$$w_{n+1} = \sum_{j=0}^{n} a_{n-j}e^{-\bar{y}_j}(1 - e^{-w_j}), n = 0, 1, \ldots. \tag{4.4.10}$$

From Lemma 4.4.2 it follows that $w_n \geq 0$. To complete the proof of the theorem it suffices to show that the zero solution of Eq.(4.4.10) is globally asymptotically stable.

Consider the function V defined by

$$V(w_n, n) = (1 - a)^{-1}\left(w_n + \sum_{r=0}^{n-1} \sum_{s=n}^{\infty} a_{s-r}w_r\right). \tag{4.4.11}$$

Since for each $r = 0, \ldots, n - 1$ the series $\sum_{s=n}^{\infty} a_{s-r}$ converges and because $\{w_n\}$ is nonnegative sequence, it follows that for every integer $n \geq 0$ the function V is well-defined and nonnegative. The function V plays the role of a "Liapunov function".

Since $(1 - a)^{-1} > 1$, it is easy to see that for all integers n,

$$V(w_n, n) > w_n. \tag{4.4.12}$$

Next we prove that for every positive solution $\{w_n\}$ of Eq.(4.4.10),

$$\Delta V(w_n, n) = V(w_{n+1}, n + 1) - V(w_n, n) \leq -w_n, \quad n = 0, 1, \ldots. \tag{4.4.13}$$

Using the facts that $\{w_n\}$ is a positive solution of Eq.(4.4.10), $\bar{y}_n \geq 0$, and $1 - e^w \leq w$ for $w \geq 0$, we find

$$
\begin{aligned}
\Delta V(w_n, n) &= V(w_{n+1}, n + 1) - V(w_n, n) \\
&= (1 - a)^{-1}(w_{n+1} + \sum_{r=0}^{n} \sum_{s=n+1}^{\infty} a_{s-r} w_r) \\
&\quad - (1 - a)^{-1}(w_n + \sum_{r=0}^{n-1} \sum_{s=n}^{\infty} a_{s-r} w_r) \\
&= (1 - a)^{-1}(\sum_{j=0}^{n} a_{n-j} e^{-\bar{y}_j}(1 - e^{-w_j}) + \sum_{r=0}^{n} \sum_{s=n+1}^{\infty} a_{s-r} w_r) \\
&\quad - (1 - a)^{-1}(w_n + \sum_{r=0}^{n-1} \sum_{s=n}^{\infty} a_{s-r} w_r) \\
&\leq (1 - a)^{-1}(\sum_{j=0}^{n} a_{n-j} w_j + \sum_{r=0}^{n-1} \sum_{s=n+1}^{\infty} a_{s-r} w_r \\
&\quad + \sum_{s=n+1}^{\infty} a_{s-n} w_n - y_n - \sum_{r=0}^{n-1} \sum_{s=n+1}^{\infty} a_{s-r} w_r - \sum_{r=0}^{n-1} a_{n-r} w_r) \\
&= (1 - a)^{-1}(a_0 y_n + \sum_{s=n+1}^{\infty} a_{s-n} w_n - w_n) \\
&= (1 - a)^{-1}(\sum_{s=n}^{\infty} a_{s-n} - 1) w_n.
\end{aligned}
$$

Since $\sum_{s=n}^{\infty} a_{s-n} = a$ it follows from the above that

$$\Delta V(w_n, n) \leq (1 - a)^{-1}(a - 1) w_n = -w_n$$

and (4.4.13) is proved.

From (4.4.13) it follows that the sequence $\{V(w_n, n)\}$ is nonincreasing for all nonnegative solutions $\{w_n\}$ of Eq.(4.4.10) and so $\{V(w_n, n)\}$ is convergent. Letting $n \to \infty$ into (4.4.13) we obtain

$$\lim_{n \to \infty} \Delta V(w_n, n) = 0 \leq - \lim_{n \to \infty} w_n$$

which implies,

$$\lim_{n \to \infty} w_n = 0.$$

To complete the proof of the theorem it remains to establish the local stability of the zero solution. From (4.4.12) and (4.4.13) it follows that

$$
\begin{aligned}
w_n &\leq V(w_n, n) \leq V(w_1, 1) \\
&= (1-a)^{-1}\left(w_0 + \sum_{s=0}^{\infty} a_s w_0\right) \\
&\leq \frac{1+a}{1-a} w_0
\end{aligned}
$$

from which the stability of the zero solution of Eq.(4.4.10) follows. The proof is complete.   □

Note that when $\{a_n\}$ is given by (4.4.6) and $a \in (0,1)$, condition (4.4.2) becomes

$$\frac{c}{1-a} < 1$$

which is equivalent to $a + c < 1$.

Open Problem 4.4.1 A Model of the Spread of an Epidemic I *Let $\{y_n\}$ and $\{z_n\}$ be solutions of Eq.(4.4.3) and Eq.(4.4.4), respectively, such that*

$$|y_0 - z_0| < \delta$$

where δ is a given positive number. Estimate, under appropriate conditions, in terms of δ the difference

$$|y_n - z_n| \quad \text{for} \quad n = 1, 2, \ldots.$$

Open Problem 4.4.2 A Model of the Spread of an Epidemic II *Study the oscillation, the stability, and the existence of periodic solutions of Eq.(4.4.3) when*

$$\sum_{n=0}^{\infty} a_n = \infty.$$

4.5   Nicholson's Blowflies

Our aim in this section is to investigate the oscillation and the global attractivity of the delay difference equation

$$N_{n+1} - N_n = -\delta N_n + P N_{n-k} e^{-a N_{n-k}}, \ n = 0, 1, \ldots \qquad (4.5.1)$$

where k is a nonnegative integer and

$$a \in (0, \infty), \ \delta \in (0, 1), \ \text{and} \ P \in (\delta, \infty). \qquad (4.5.2)$$

Eq.(4.5.1) is a discrete analogue of the delay differential equation

$$\dot{N}(t) = -\delta N(t) + P N(t - \tau) e^{-a N(t-\tau)}, \ t \geq 0 \qquad (4.5.3)$$

which has been used in describing the dynamics of Nicholson's blowflies. See Gurney, Blythe and Nisbet [1], Györi and Ladas [1,2], Kocic and Ladas [1], Kulenović, Ladas and Sficas [1] and Nicholson [1].

It is easy to see that when (4.5.2) holds and the initial conditions are such that

$$N_{-k}, N_{-k+1}, \ldots, N_{-1} \in [0, \infty) \ \text{and} \ N_0 \in (0, \infty) \qquad (4.5.4)$$

then the corresponding solution $\{N_n\}$ is positive. Furthermore the unique positive equilibrium $\bar{N}$ of Eq.(4.5.1) is given by

$$\bar{N} = \frac{1}{a} \ln \left(\frac{P}{\delta} \right). \qquad (4.5.5)$$

Theorem 4.5.1 *Assume that (4.5.2) holds and that*

$$\delta \left(\ln \frac{P}{\delta} - 1 \right) \frac{(k+1)^{k+1}}{k^k} > (1 - \delta)^{k+1}. \qquad (4.5.6)$$

Then every positive solution of Eq.(4.5.1) oscillates about the positive equilibrium $\bar{N}$.

Proof. Assume for the sake of contradiction that Eq.(4.5.1) has a positive solution $\{N_n\}$ which does not oscillate about $\bar{N}$. Set

$$N_n = \bar{N} + \frac{1}{a} x_n, \ n = -k, -k+1, \ldots. \qquad (4.5.7)$$

Then $\{x_n\}$ is a nonoscillatory solution of the difference equation

$$x_{n+1} - x_n + \delta x_n + a\delta \bar{N}(1 - e^{-x_{n-k}}) - \delta x_{n-k} e^{-x_{n-k}} = 0, n = 0, 1, \ldots. \qquad (4.5.8)$$

We will assume that $\{x_n\}$ is eventually positive. The case where $\{x_n\}$ is eventually negative is similar and will be omitted. Let n_0 be an integer such that

$$x_n > 0 \quad \text{for} \quad n \geq n_0. \tag{4.5.9}$$

First we claim that $\{x_n\}$ is a bounded sequence. Otherwise there exists a subsequence $\{x_{n_i}\}$ such that for $i = 1, 2, \ldots$

$$n_i \geq n_0, \ \lim_{i \to \infty} x_{n_i} = \infty \quad \text{and} \quad x_{n_i+1} - x_{n_i} \geq 0.$$

It follows from (4.5.8) that for i sufficiently large

$$x_{n_i} + a\bar{N} \leq (x_{n_i-k} + a\bar{N})e^{-x_{n_i-k}} \tag{4.5.10}$$
$$\leq x_{n_i-k} + a\bar{N}. \tag{4.5.11}$$

From (4.5.11) we see that $\lim_{i \to \infty} x_{n_i} = \infty$. But then (4.5.10) leads to a contradiction, as $i \to \infty$.

Next we claim that

$$\lim_{n \to \infty} x_n = 0. \tag{4.5.12}$$

Otherwise, let

$$\mu = \limsup_{n \to \infty} x_n. \tag{4.5.13}$$

Then $\mu > 0$ and there exists a subsequence $\{x_{n_i}\}$ such that for $i = 1, 2, \ldots$

$$n_i \geq n_0, \ \lim_{i \to \infty} x_{n_i} = \mu \quad \text{and} \quad x_{n_i+1} - x_{n_i} \geq 0.$$

Also (4.5.10) and (4.5.11) hold. From (4.5.11) we see that

$$\mu \leq \limsup_{i \to \infty} x_{n_i-k}$$

and so because of (4.5.13),

$$\limsup_{i \to \infty} x_{n_i-k} = \mu.$$

But then (4.5.10) leads to

$$\mu + a\bar{N} \leq (\mu + a\bar{N})e^{-\mu} < \mu + a\bar{N}$$

which is impossible. Hence (4.5.12) holds.

Eq.(4.5.8) can be rewritten in the form

$$x_{n+1} - x_n + \delta x_n + P(n)x_{n-k} = 0, \ n = 0, 1, \ldots \tag{4.5.14}$$

where

$$P(n) = a\delta\bar{N}\frac{1 - e^{x_{n-k}}}{x_{n-k}} - \delta x_{n-k}$$

and

$$\lim_{n\to\infty} P(n) = a\delta\bar{N} - \delta = \delta\left(\ln\frac{P}{\delta} - 1\right).$$

One can easily see that the hypotheses of Theorem 1.2.2 are satisfied and so the linear equation

$$y_{n+1} - y_n + \delta y_n + \delta\left(\ln\frac{P}{\delta} - 1\right)y_{n-k} = 0, \quad n = 0, 1, \ldots \tag{4.5.15}$$

has an eventually positive solution. Let $\{y_n\}$ be an eventually positive solution of Eq.(4.5.15). Then $z_n = (1 - \delta)^n y^n$ is an eventually positive solution of

$$z_{n+1} - z_n + \delta(1 - \delta)^{-k-1}\left(\ln\frac{P}{\delta} - 1\right)z_{n-k} = 0, \quad n = 0, 1, \ldots. \tag{4.5.16}$$

According to Corollary 1.2.1, Eq.(4.5.16) has no nonoscillatory solutions and this contradiction completes the proof. □

Theorem 4.5.2 *Assume that (4.5.2) holds and that*

$$[(1 - \delta)^{-k-1} - 1]\left(\frac{P}{\delta} - 1\right) < 1. \tag{4.5.17}$$

Then the positive equilibrium $\bar{N}$ of Eq.(4.5.1) is globally asymptotically stable.

Proof. First we will prove that the positive equilibrium $\bar{N}$ is a global attractor of all positive solutions of Eq.(4.5.1). The proof will be accomplished by introducing the transformation (4.5.7) and then by showing that (4.5.12) is satisfied. A slight modification of Theorem 4.5.1 shows that (4.5.12) is true when eventually $x_n \geq 0$ or when eventually $x_n \leq 0$. Therefore it remains to establish (4.5.12) when $\{x_n\}$ is strictly oscillatory. To this end, let

$$\{x_{p_i+1}, x_{p_i+2}, \ldots, x_{q_i}\}$$

be the ith positive semicycle of $\{x_n\}$ followed by the ith negative semicycle

$$\{x_{q_i+1}, x_{q_i+2}, \ldots, x_{p_{i+1}}\}.$$

Let x_{M_i} and x_{m_i} be the extreme values in these two semicycles, respectively, with the smallest possible indices M_i and m_i. Then we claim that

$$M_i - p_i \leq k + 1 \quad \text{and} \quad m_i - q_i \leq k + 1. \tag{4.5.18}$$

We will prove (4.5.18) for positive semicycles. The proof for negative semicycles is similar and will be omitted. Assume for the sake of contradiction that the first inequality in (4.5.18) is not true. Then $M_i - p_i > k+1$ and the terms $x_{M_i-1}, \ldots, x_{M_i-k-1}$ are in a positive semicycle. Since $x_{M_i} \geq x_{M_i-1}$, from Eq.(4.5.12) we obtain

$$x_{M_i-1} + a\bar{N} \leq (x_{M_i-k-1} + a\bar{N})e^{-x_{M_i-k-1}} < x_{M_i-k-1} + a\bar{N}.$$

Furthermore, from (4.5.12) and the above we find

$$x_{M_i} + a\bar{N} = (1-\delta)(x_{M_i-1} + a\bar{N}) + \delta(x_{M_i-k-1} + a\bar{N})e^{-x_{M_i-k-1}}$$

$$< (1-\delta)(x_{M_i-k-1} + a\bar{N}) + \delta(x_{M_i-k-1} + a\bar{N})$$

$$= x_{M_i-k-1} + a\bar{N}.$$

This is a contradiction and so (4.5.18) is true.

Let

$$\left.\begin{array}{rcl} \lambda & = & \liminf\limits_{n\to\infty} x_n = \liminf\limits_{i\to\infty} x_{m_i} \\[2mm] \mu & = & \limsup\limits_{n\to\infty} x_n = \limsup\limits_{i\to\infty} x_{M_i}. \end{array}\right\} \tag{4.5.19}$$

Clearly $\lambda \geq -a\bar{N}$. To prove that (4.5.12) holds it is sufficient to show that

$$\lambda = \mu = 0. \tag{4.5.20}$$

From (4.5.19) it follows that if $\eta \in (0,\infty)$ and $\epsilon \in (0,\lambda)$ are given, then there exists $n_0 \in \mathbb{N}$ such that

$$\lambda - \epsilon \leq x_n \leq \mu \quad \text{for} \quad n \geq n_0 + k. \tag{4.5.21}$$

Furthermore we have

$$x_{n-k}e^{-x_{n-k}} \leq \mu + \epsilon \quad \text{for} \quad n \geq n_0 + k. \tag{4.5.22}$$

Eq. (4.5.8) can be written in the form

$$x_{n+1} - (1-\delta)x_n = -\delta a\bar{N} + \delta a\bar{N}e^{-x_{n-k}} + \delta x_{n-k}e^{-x_{n-k}}. \tag{4.5.23}$$

By multiplying Eq.(4.5.23) by $(1-\delta)^{-n-1}$ and then summing up from $n = p_i$ to $n = M_i - 1$, for i sufficiently large, we obtain

$$(1-\delta)^{-M_i}x_{M_i} - (1-\delta)^{-p_i}x_{p_i} = -\delta a\bar{N}\sum_{j=p_i}^{M_i-1}(1-\delta)^{-j-1}$$

$$+ \quad \delta a \bar{N} \sum_{j=p_i}^{M_i-1} e^{-x_{j-k}}(1-\delta)^{-j-1}$$

$$+ \quad \delta \sum_{j=p_i}^{M_i-1} x_{j-k} e^{-x_{j-k}}(1-\delta)^{-j-1}.$$

Since $x_{p_i} \le 0$ and $e^{-x_{n-k}} < e^{a\bar{N}}$, it follows from (4.5.22) and a simple calculation that

$$x_{M_i} \le (a\bar{N}(e^{a\bar{N}} - 1) + \mu + \epsilon)[1 - (1-\delta)^{M_i-p_i}].$$

As $M_i - p_i \le k+1$ and $a\bar{N} = \ln(P/\delta)$, we find

$$x_{M_i} \le (a\bar{N}(e^{a\bar{N}} - 1) + \mu + \epsilon)[1 - (1-\delta)^{k+1}].$$

By using (4.5.19) we obtain

$$\mu \le (a\bar{N}(e^{a\bar{N}} - 1) + \mu + \epsilon)[1 - (1-\delta)^{k+1}].$$

As $\epsilon > 0$ is arbitrary, we have

$$\mu \le [a\bar{N}(\frac{P}{\delta} - 1) + \mu][(1-\delta)^{-k-1} - 1].$$

Set

$$Q_1 = [a\bar{N}(\frac{P}{\delta} - 1) + \mu][(1-\delta)^{-k-1} - 1]. \tag{4.5.24}$$

Then

$$\mu \le Q_1, \quad 0 < Q_1 < a\bar{N}, \quad \text{and} \quad e^{Q_1} < e^{a\bar{N}} = \frac{P}{\delta}. \tag{4.5.25}$$

Furthermore from (4.5.19), it follows that for n sufficiently large and for $\epsilon \ge 0$ and arbitrarily small,

$$x_{n-k} + a\bar{N} \ge \lambda + a\bar{N} - \epsilon \ge 0.$$

Also from (4.5.25) for $\epsilon' > 0$ and n sufficiently large we have

$$x_{n-k} \le \mu + \epsilon'; \quad \text{that is,} \quad e^{-x_{n-k}} \ge e^{-Q_1-\epsilon'}$$

and

$$(a\bar{N} + x_{n-k})e^{-x_{n-k}} \ge (a\bar{N} + \lambda - \epsilon')e^{-Q_1-\epsilon'}. \tag{4.5.26}$$

After multiplying Eq.(4.5.23) by $(1-\delta)^{-n-1}$ and then summing up from $n = q_i$ to $n = m_i - 1$, for i sufficiently large, we obtain

$$(1-\delta)^{-m_i} x_{m_i} - (1-\delta)^{-q_i} x_{q_i} = -\delta a\bar{N} \sum_{j=q_i}^{m_i-1} (1-\delta)^{-j-1}$$

$$+ \quad \delta \sum_{j=q_i}^{m_i-1} (a\bar{N} + x_{j-k} e^{-x_{j-k}})(1-\delta)^{-j-1}.$$

Using (4.5.26) and (4.5.19), and since $x_{q_i} \geq 0$, $x_{m_i} < 0$, $m_i - q_i \leq k+1$, and $\epsilon \geq 0$, $\epsilon' > 0$ are arbitrary, we have

$$0 \leq -\lambda \leq [a\bar{N}(1 - e^{Q_1}) - \lambda e^{-Q_1}][1 - (1 - \delta)^{k+1}].$$

That is,

$$-\lambda[1 - e^{Q_1}][1 - (1 - \delta)^{k+1}] \leq a\bar{N}(1 - e^{Q_1})[1 - (1 - \delta)^{k+1}].$$

It is easy to see that

$$[1 - e^{Q_1}][1 - (1 - \delta)^{k+1}] \geq e^{-Q_1}(1 - \delta)^{k+1}$$

and therefore

$$- \lambda \leq a\bar{N}(1 - e^{Q_1})[(1 - \delta)^{-k-1} - 1] \leq a\bar{N}(\frac{P}{\delta} - 1)[(1 - \delta)^{-k-1} - 1] = Q_1. \quad (4.5.27)$$

Hence

$$- Q_1 \leq \lambda \leq \mu \leq Q_1 \quad \text{and} \quad 0 < Q_1 < a\bar{N}. \quad (4.5.28)$$

If we define the sequence $\{Q_n\}$ by

$$Q_{n+1} = a\bar{N}[(1 - \delta)^{-k-1} - 1](e^{Q_n} - 1) \quad (4.5.29)$$

where Q_1 is given by (4.5.24), then from (4.5.27) we find

$$Q_2 \leq Q_1.$$

It easy to prove by induction that for $n = 1, 2, \ldots$

$$-Q_n \leq \lambda \leq \mu \leq Q_n, \quad 0 < Q_n < a\bar{N}, \quad \text{and} \quad Q_{n+1} \leq Q_n.$$

In particular, $\{Q_n\}$ is a convergent sequence. Set

$$Q = \lim_{n \to \infty} Q_n.$$

Then Q is a solution of the equation

$$Q = a\bar{N}[(1 - \delta)^{-k-1} - 1](e^Q - 1)$$

on the interval $[0, a\bar{N}]$. Since the function F, which is defined by

$$F(u) = a\bar{N}[(1 - \delta)^{-k-1} - 1](e^u - 1) - u$$

has two zeros on $[0, \infty)$, but only one on $[0, a\bar{N}]$, we conclude that $Q = 0$. This implies that

$$\lambda = \mu = 0$$

and the proof is complete. $\square$

Research Project 4.5.1 *Study the oscillations and global stability of some discrete analogue of the delay differential equation*

$$\frac{dP(t)}{dt} = -\alpha P(t) + \frac{\beta P(t-\tau)}{1 + [P(t-\tau)]^n}, \; t \geq 0$$

which has been proposed by Mackey and Glass [1] as a model of haematopoiesis. See also Györi and Ladas [2].

4.6 A Discrete Analogue of a Model of Haematopoiesis

Our aim in this section is to investigate the positive solutions of the difference equation

$$N_{n+1} = \alpha N_n + \frac{\beta}{1 + N_{n-k}^p}, \; n = 0, 1, \ldots \tag{4.6.1}$$

where

$$\alpha \in [0, 1), \quad p, \beta \in (0, \infty), \quad k \in \{1, 2, \ldots\} \tag{4.6.2}$$

and where the initial conditions $N_{-k}, \ldots, N_0$ are nonnegative.

Eq.(4.6.1) is a discrete analogue of of the delay differential equation

$$\frac{dP(t)}{dt} = \frac{\beta_0 \theta^n}{\theta^n + P(t-\tau)^n} - \gamma P(t), \; t \geq 0$$

which has been proposed by Mackey and Glass [1] as a model of haematopoiesis (blood cell production). See also Karakostas, Philos and Sficas [1]. Here β_0, θ, γ, τ and n are positive constants and $P(t)$ denotes the density of mature cells in blood circulation.

Eq.(4.6.1) has a unique positive equilibrium $\bar{N}$. Furthermore, $\bar{N}$ is a solution of the equation

$$\bar{N} + \bar{N}^{p+1} = \frac{\beta}{1-\alpha}. \tag{4.6.3}$$

The following result which is a direct consequence of Corollary 2.4.1 is very useful in studying the global attractivity of Eq.(4.6.1).

Theorem 4.6.1 *Assume that (4.6.2) holds. Let $\bar{x}$ be the unique positive equilibrium of the first order difference equation*

$$x_{n+1} = 1 + \frac{A}{x_n^p}, \; n = 0, 1, 2, \ldots \tag{4.6.4}$$

where p and x_0 are positive numbers and

$$A = \left(\frac{\beta}{1-\alpha} \right)^p. \tag{4.6.5}$$

If $\bar{x}$ is a global attractor of all positive solutions of Eq.(4.6.4), then $\bar{N}$ is a global attractor of all positive solutions of Eq.(4.6.1).

Proof. We will apply Corollary 2.4.1. It is easy to see that the function

$$F(u) = \frac{\beta}{1 + u^p}$$

satisfies all the hypotheses of Corollary 2.4.1. Consider the difference equation

$$y_{n+1} = \frac{\beta}{(1-\alpha)(1 + y_n^p)}, \ n = 0, 1, \ldots. \tag{4.6.6}$$

If $\bar{N}$ is a global attractor of all positive solutions of Eq.(4.6.6), then $\bar{N}$ is also a global attractor of all positive solutions of Eq.(4.6.1). On the other hand, the change of variables

$$y_n = \frac{\beta}{(1-\alpha)x_n}$$

transforms Eq.(4.6.6) to Eq.(4.6.4) from which the result follows.   □

Remark 4.6.1 Eq.(4.6.4) has a unique positive equilibrium $\bar{x}$. Furthermore $\bar{x}$ is the unique positive root of the equation

$$\bar{x}^{p+1} - \bar{x}^p - A = 0. \tag{4.6.7}$$

Theorem 4.6.1 shows that the behavior of solutions of Eq.(4.6.1) is closely related to the behavior of solutions of the first-order equation (4.6.4). The following result from DeVault, Kocic and Ladas [1] summarizes all known properties of the solutions of Eq.(4.6.4).

Theorem 4.6.2 *Consider the difference equation (4.6.4) where A , p , and x_0 are positive numbers. Then the following statements are true.*

(a) The positive equilibrium $\bar{x}$ of Eq.(4.6.4) is globally asymptotically stable if and only if

$$p \le 1 \tag{4.6.8}$$

or

$$p > 1 \ \text{and} \ A \le \frac{p^p}{(p-1)^{p+1}}. \tag{4.6.9}$$

(b) When

$$p > 1 \quad \text{and} \quad A > \frac{p^p}{(p-1)^{p+1}} \tag{4.6.10}$$

Eq.(4.6.4) has two periodic solutions, each of period two, of the form $S_1 : \bar{x}_1, \bar{x}_2,$ $\bar{x}_1, \bar{x}_2, \ldots$ and $S_2 : \bar{x}_2, \bar{x}_1, \bar{x}_2, \bar{x}_1, \ldots$ which are asymptotically stable and global attractors with basin of attraction $(0, \bar{x})$ and $(\bar{x}, \infty)$, respectively.

(c) The subsequences $\{x_{2n}\}$ and $\{x_{2n+1}\}$ of every solution of Eq.(4.6.4) converge monotonically to $\bar{x}$ in case (a), and to S_1 or S_2 in case (b).

(d) Every solution of Eq.(4.6.4) oscillates about every other solution of the same equation.

Proof. The proof of the theorem will be established in the following series of five lemmas.

Lemma 4.6.1 Assume that A , p , and x_0 are positive numbers. Then the positive equilibrium $\bar{x}$ of Eq.(4.6.4) is locally asymptotically stable if and only if one of the conditions (4.6.8) or (4.6.9) is satisfied.

Proof. Using linearized stability we see that if $p \le 1$, $\bar{x}$ is locally asymptotically stable. On the other hand, if $p > 1$, $\bar{x}$ is locally asymptotically stable provided that $\bar{x} < p/(p-1)$, and is unstable if $\bar{x} > p/(p-1)$. This condition can be seen to be equivalent to

$$A < \frac{p^p}{(p-1)^{p+1}}.$$

When

$$A = \frac{p^p}{(p-1)^{p+1}}$$

by using the Schwarzian derivative we see that $\bar{x}$ is locally asymptotically stable. $\square$

Lemma 4.6.2 Assume that A and x_0 are positive numbers and that $0 < p \le 1$. Then the positive equilibrium $\bar{x}$ of Eq.(4.6.4) is globally asymptotically stable.

Proof. In the proof we will apply Lemma 1.6.5. For Eq.(4.6.4) the associated algebraic system is

$$U = 1 + \frac{A}{L^p} , \quad L = 1 + \frac{A}{U^p}. \tag{4.6.11}$$

Multiplying the first equation of this system by L and the second by U gives

$$UL = L + AL^{1-p} , \quad UL = U + AU^{1-p}$$

and so

$$0 = (U - L) + A(U^{1-p} - L^{1-p}).$$

For $p \leq 1$, this implies that $U = L = \bar{x}$ and so, by Lemmas 1.6.5 and 4.6.1, $\bar{x}$ is globally asymptotically stable, with the subsequences $\{x_{2n}\}$ and $\{x_{2n+1}\}$ converging monotonically.   □

Lemma 4.6.3 *Assume that $p \in (0,1)$, A and x_0 are positive numbers, and that*

$$A \leq \frac{p^p}{(p-1)^{p+1}}.$$

Then the positive equilibrium $\bar{x}$ of Eq.(4.6.4) is globally asymptotically stable.

Proof. Once again it must be shown that the system (4.6.11) has the unique solution $U = L = \bar{x}$. Let

$$F(u) = 1 + \frac{A}{\left(1 + \dfrac{A}{u^p}\right)^p}, \quad u > 0$$

where $F(u) = f(f(u))$ is the second iterate of $f(u) = 1 + \frac{A}{u^p}$. Then

$$F(\bar{x}) = 1 + \frac{A}{\left(1 + \dfrac{A}{\bar{x}^p}\right)^p} = 1 + \frac{A}{\bar{x}^p} = \bar{x}.$$

Thus $\bar{x}$ is a solution of $F(u) = u$ and hence $L = U = \bar{x}$ is a solution of (4.6.11). We will show that this is the only solution of the system. Set

$$H(u) = F(u) - u, \quad u > 0.$$

Then $\bar{x}$ is a zero of $H(u)$. It suffices to show that $H(u)$ is strictly decreasing. To this end observe that

$$H'(u) = \frac{(Ap)^2}{\left(u + \dfrac{A}{u^{p-1}}\right)^{p+1}} - 1 = \frac{(Ap)^2}{G(u)^{p+1}} - 1$$

where

$$G(u) = u + \frac{A}{u^{p-1}} = u\left(1 + \frac{A}{u^p}\right).$$

Since the function $G(u)$ has a minimum at $u = u_0 = ((p-1)A)^{\frac{1}{p}}$,

$$G(u) \geq G(u_0) = u_0\left(1 + \frac{A}{u_0^p}\right)$$

and so
$$G(u) \geq \frac{p}{p-1}((p-1)A)^{\frac{1}{p}}$$

with equality holding only when $u = u_0$. Now

$$H'(u) \ \leq \ \frac{(Ap)^2}{G(u_0)^{p+1}} - 1 = \frac{(Ap)^2}{((p-1)A)^{\frac{p+1}{p}}(\frac{p}{p-1})^{p+1}} - 1$$

$$\leq \ \left(\frac{A(p-1)^{p+1}}{p^p}\right)^{\frac{p-1}{p}} - 1.$$

Thus if

$$A \leq \frac{p^p}{(p-1)^{p+1}}$$

then $H'(u) \leq 0$, and since equality holds only if $u = u_0$ and $A = p^p/(p-1)^{p+1}$, it follows that $H(u)$ is strictly decreasing. Therefore $H(u)$ has only one zero and this is the unique positive equilibrium $\bar{x}$. Hence the system (4.6.11) has only one solution, namely $U = L = \bar{x}$, and so $\bar{x}$ is a global attractor of Eq.(4.6.4). $\qquad\square$

Lemma 4.6.4 *Assume that A and x_0 are positive numbers and that (4.6.10) holds. Then Eq.(4.6.4) has two periodic solutions, each of period two, of the form $S_1 : \bar{x}_1, \bar{x}_2, \bar{x}_1, \bar{x}_2, \ldots$ and $S_2 : \bar{x}_2, \bar{x}_1, \bar{x}_2, \bar{x}_1, \ldots.$*

Proof. We will show that the equation $H(u) = 0$ has exactly three roots $\bar{x}_1, \bar{x}, \bar{x}_2$. Thus the system (4.6.11) has the three solutions: $\{\bar{x}, \bar{x}\}$, $\{\bar{x}_1, \bar{x}_2\}$, and $\{\bar{x}_2, \bar{x}_1\}$. As in Lemma 4.6.3,

$$H'(u) = \frac{(Ap)^2}{G(u)^{p+1}} - 1$$

where $H'(u)$ obtains a global maximum at $u = u_0 = (A(p-1))^{\frac{1}{p}}$. Therefore

$$0 < H'(u_0) = \left(\frac{A(p-1)^{p+1}}{p^p}\right)^{\frac{p-1}{p}} - 1 < \frac{(Ap)^2}{G(u_0)^{p+1}} - 1$$

and so $G(u_0) < (Ap)^{2/(p+1)}$. Clearly $G(u)$ is decreasing on $(0, u_0)$ and increasing on (u_0, ∞). Thus there exist exactly two solutions u_1 and u_2, with $0 < u_1 < u_0 < u_2$, of the equation $G(u) = (Ap)^{2/(p+1)}$ which gives $H'(u_1) = H'(u_2) = 0$. Now since $G(u) > (Ap)^{2/(p+1)}$ for $u \in (0, u_1) \cup (u_2, \infty)$, and $G(u) < (Ap)^{2/(p+1)}$ for $u \in (u_1, u_2)$, it follows that $H(u)$ is decreasing on $(0, u_1) \cup (u_2, \infty)$ and increasing on (u_1, u_2). This means that $H(u)$ has no more than three zeros for $u > 0$, since if there exist more than three zeros, then by Rolles Theorem there must exist more than two points such that $H'(u) = 0$.

We will now show that there exist exactly three roots of the equation $H(u) = 0$. As $H(1) = A/(1 + A)^p > 0$, $H(\infty) = -\infty$, and $H(\bar{x}) = 0$, it suffices to show that $H'(\bar{x}) > 0$, since then there exists $\delta > 0$ such that $H(x) < 0$ for $x \in (\bar{x} - \delta, \bar{x})$ and $H(x) > 0$ for $x \in (\bar{x}, \bar{x} + \delta)$, and so there exist zeros of $H(u)$ in $(1, \bar{x})$ and $(\bar{x}, \infty)$. Let

$$h(u) = 1 + \frac{A}{u^p} - u , \ u > 0.$$

Then

$$h'(u) = \frac{-Ap}{u^{p+1}} - 1 < 0 , \ u > 0$$

and so $h(u)$ is decreasing and $h(\bar{x}) = 0$. Hence $h(u) > 0$ for $u \in (0, \bar{x})$, $h(u) < 0$ for $u \in (\bar{x}, \infty)$, and

$$H'(\bar{x}) = \frac{(Ap)^2}{\left(1 + \dfrac{A}{\bar{x}^p}\right)^{p+1} \bar{x}^{p+1}} - 1 = \frac{(Ap)^2}{\bar{x}^{2(p+1)}} - 1.$$

Thus $H'(\bar{x}) > 0$ if $\bar{x} < (Ap)^{1/(p+1)}$. Next we will show that $\bar{x} < (Ap)^{1/(p+1)}$. Observe that

$$h((Ap)^{\frac{1}{p+1}}) = 1 + \frac{A}{(Ap)^{\frac{p}{p+1}}} - (Ap)^{\frac{1}{p+1}} = 1 - \left(\frac{A(p-1)^{p+1}}{p^p}\right)^{\frac{1}{p+1}} < 0$$

because $A > p^p/(p-1)^{p+1}$. Therefore $\bar{x} < (Ap)^{1/(p+1)}$ which implies that $H'(\bar{x}) > 0$, and so it follows $\bar{x}_1$, $\bar{x}$, and $\bar{x}_2$ are the only zeros of $H(u) = 0$. Clearly

$$\bar{x}_1 = 1 + \frac{A}{\bar{x}_2^p} , \quad \bar{x}_2 = 1 + \frac{A}{\bar{x}_1^p}$$

and so (4.6.11) has exactly three solutions in $(0, \infty) \times (0, \infty)$ namely, $\{\bar{x}, \bar{x}\}$, $\{\bar{x}_1, \bar{x}_2\}$, and $\{\bar{x}_2, \bar{x}_1\}$. So if $A > p^p/(p-1)^{p+1}$, Eq.(4.6.4) has two solutions of period two which are $(\bar{x}_1, \bar{x}_2)$ and $(\bar{x}_2, \bar{x}_1)$.   □

Lemma 4.6.5 *Assume that A and x_0 are positive numbers and (4.6.10) holds. Then the sequence $\{x_n\}$ diverges; if $\{\bar{x}_1, \bar{x}_2\}, \{\bar{x}, \bar{x}\}$, and $\{\bar{x}_2, \bar{x}_1\}$ with $\bar{x}_1 < \bar{x} < \bar{x}_2$ are the three solutions of the system (4.6.11), then the following statements are true:*

(a) *If $0 < x_0 < \bar{x}_1$, then $\{x_{2n}\}$ is increasing, $\{x_{2n+1}\}$ is decreasing, $\lim_{n\to\infty} x_{2n} = \bar{x}_1$, and $\lim_{n\to\infty} x_{2n+1} = \bar{x}_2$.*

(b) *If $\bar{x}_1 < x_0 < \bar{x}$, then $\{x_{2n}\}$ is decreasing, $\{x_{2n+1}\}$ is increasing, $\lim_{n\to\infty} x_{2n} = \bar{x}_1$, and $\lim_{n\to\infty} x_{2n+1} = \bar{x}_2$.*

(c) *If $\bar{x} < x_0 < \bar{x}_2$, then $\{x_{2n}\}$ is increasing, $\{x_{2n+1}\}$ is decreasing, $\lim_{n\to\infty} x_{2n} = \bar{x}_2$, and $\lim_{n\to\infty} x_{2n+1} = \bar{x}_1$.*

(d) *If $\bar{x}_2 < x_0 < \infty$, then $\{x_{2n}\}$ is decreasing, $\{x_{2n+1}\}$ is increasing, $\lim_{n\to\infty} x_{2n} = \bar{x}_2$, and $\lim_{n\to\infty} x_{2n+1} = \bar{x}_1$.*

(e) *If $x_0 = \bar{x}_1$, then $x_{2n} = \bar{x}_1$, and $x_{2n+1} = \bar{x}_2$ for $n = 0, 1, \ldots$.*

(f) *If $x_0 = \bar{x}_2$, then $x_{2n} = \bar{x}_2$, and $x_{2n+1} = \bar{x}_1$ for $n = 0, 1, \ldots$.*

Proof. Indeed, (e) and (f) have already been proved. Let $f(x) = 1 + A/x^p$. If $0 < u < \bar{x}_1$ and $v > \bar{x}_2$ then

$$f(u) > f(\bar{x}_1) = \bar{x}_2 \quad \text{and} \quad f(v) < f(\bar{x}_2) = \bar{x}_1$$

and so cases (a) and (d) are dual. Similarly if $\bar{x}_1 < u < \bar{x}$ and $\bar{x} < v < \bar{x}_2$, then

$$f(u) < f(\bar{x}_1) = \bar{x}_2, \qquad f(u) > f(\bar{x}) = \bar{x},$$
$$f(v) > f(\bar{x}_2) = \bar{x}_1, \qquad f(v) < f(\bar{x}) = \bar{x}$$

and so cases (b) and (c) are dual. Thus it suffices to consider only (a) and (b). We will prove (a). The proof of (b) is similar and will be omitted. As before notice that for the function $H(u)$, we have $H(u) > 0$ for $u \in (0, \bar{x}_1) \cup (\bar{x}, \bar{x}_2)$ and $H(u) < 0$ for $u \in (\bar{x}_1, \bar{x}) \cup (\bar{x}_2, \infty)$. This means $F(u) > u$ in the first case and $F(u) < u$ in the second case. In case (a) it is assumed that $0 < x_0 < \bar{x}_1$, giving $x_1 > \bar{x}_2$, because

$$x_1 = 1 + \frac{A}{x_0^p} > 1 + \frac{A}{\bar{x}_1^p} = \bar{x}_2.$$

Also, $x_0 < x_2 < \bar{x}_1$ since

$$x_2 = 1 + \frac{A}{x_1^p} < 1 + \frac{A}{\bar{x}_2^p} = \bar{x}_1$$

and $x_2 = F(x_0) > x_0$ since $x_0 \in (0, \bar{x}_1)$. Similarly, $\bar{x}_2 < x_3 < x_1$, because

$$x_3 = 1 + \frac{A}{x_2^p} > 1 + \frac{A}{\bar{x}_1^p} = \bar{x}_2$$

and $x_3 = F(x_1) < x_1$ since $x_1 \in (\bar{x}_2, \infty)$. Using induction we see that

$$\bar{x}_1 > x_{2n+2} = F(x_{2n}) > x_{2n} \quad \text{and} \quad \bar{x}_2 < x_{2n+3} = F(x_{2n+1}) < x_{2n+1}$$

and the sequences $\{x_{2n}\}$ and $\{x_{2n+1}\}$ are monotonic. Let

$$y = \lim_{n\to\infty} x_{2n} \quad \text{and} \quad z = \lim_{n\to\infty} x_{2n+1}.$$

Then $y = F(y)$ and $z = F(z)$ and as $y \le \bar{x}_1$ and $z \ge \bar{x}_2$, it is clear that $y = \bar{x}_1$ and $z = \bar{x}_2$. This means $\{x_{2n}\}$ increases, $\{x_{2n+1}\}$ decreases, $\lim_{n \to \infty} x_{2n} = \bar{x}_1$ and $\lim_{n \to \infty} x_{2n+1} = \bar{x}_2$. The proof of (a) is complete. □

To complete the proof Theorem 4.6.2 it remains to prove part (d). To this end, let $\{x_n\}$ and $\{y_n\}$ be solutions of Eq.(4.6.4). Then

$$x_{n+1} - y_{n+1} = \frac{A}{x_n^p} - \frac{A}{y_n^p} = A \frac{y_n^p - x_n^p}{x_n^p y_n^p}.$$

So if $x_n > y_n$, then $x_{n+1} - y_{n+1} < 0$ which implies that $y_{n+1} > x_{n+1}$. Similarly $y_n > x_n$ implies $x_{n+1} > y_{n+1}$. Thus every solution of Eq.(4.6.4) oscillates about every other solution and the proof of the Theorem is complete. □

The following Theorem deals with the global asymptotic stability of Eq.(4.6.1).

Theorem 4.6.3 *Assume that (4.6.2) holds. If*

$$\left. \begin{array}{l} \text{either} \quad p \le 1 \\[2ex] \text{or} \quad p > 1 \quad \text{and} \quad \left(\dfrac{\beta}{1-\alpha}\right)^p < \dfrac{p^p}{(p-1)^{p+1}} \end{array} \right\} \tag{4.6.12}$$

then the positive equilibrium $\bar{N}$ of Eq.(4.6.1) is globally asymptotically stable.

Proof. From Theorems 4.6.1 and 4.6.2 it follows that $\bar{N}$ is a global attractor of all positive solutions of Eq.(4.6.1). The linearized equation of Eq.(4.6.1) about the positive equilibrium $\bar{N}$ is

$$z_{n+1} - \alpha z_n + \frac{(1-\alpha)p\bar{N}^p}{1+\bar{N}^p} z_{n-k} = 0, \ n = 0, 1, \dots. \tag{4.6.13}$$

It is easy to see that if $p \le 1$ then

$$|-\alpha| + \frac{(1-\alpha)p\bar{N}^p}{1+\bar{N}^p} < \alpha + (1-\alpha) = 1$$

and according to Theorem 1.3.7, $\bar{N}$ is locally asymptotically stable. Next assume that the second inequality (4.6.12) holds. Then

$$\bar{N}(1+\bar{N}^p) < \left(\frac{1}{p-1}\right)^{1/p}\left(1+\frac{1}{p-1}\right).$$

Since the function $h(x) = x(1+x^p)$ is increasing for $x \ge 0$, it follows from the above that

$$\bar{N} < \left(\frac{1}{p-1}\right)^p.$$

Then

$$|-\alpha| + \frac{(1-\alpha)p\bar{N}^p}{1+\bar{N}^p} < \alpha + (1-\alpha) = 1$$

which completes the proof that $\bar{N}$ is locally stable. $\qquad\square$

The following result is a direct consequence of Theorem 4.6.2.

Corollary 4.6.1 *Assume that $\alpha = 0$, $p, \beta \in (0, \infty)$ and k is an even positive number. Then Eq.(4.6.1) has periodic solutions with period two.*

Research Project 4.6.1 *Study the oscillation and the global stability of the discrete analogue of the delay differential equation*

$$\frac{dN(t)}{dt} = -\alpha N(t) + \beta e^{-\gamma N(t-\tau)}, \ t \geq 0.$$

This equation has been used by Wazewska–Czyzewska and Lasota [1], see also Györi and Ladas [2], as a model for the survival of red blood cells in an animal. Here $N(t)$ denotes the number of red blood cells at time t, α is the probability of death of a red blood cell, β and γ are positive constants which are related to the production of red blood cells, and τ is the time which is required to produce a red blood cell.

4.7 A Discrete Baleen Whale Model

Our aim in this section is to investigate the global stability of the delay difference equation

$$N_{n+1} = (1-\mu)N_n + \mu N_{n-k}\left[1 + q(1 - (\frac{N_{n-k}}{K})^z)\right]_+, \ n = 0, 1, \ldots \qquad (4.7.1)$$

where k is a nonnegative integer,

$$K, q, z \in (0, \infty), \ \mu \in (0, 1) \qquad (4.7.2)$$

and where $[x]_+ = \max\{x, 0\}$. Eq.(4.7.1) has been proposed by the International Whaling Commission as a model that describes the dynamics of baleen whales. See Botsford [1], Clark [1], Fisher [1], May [4], and the references cited therein. In Eq.(4.7.1), N_n denotes the population of (sexually mature) adult whales in year n, μ is a mortality rate, K is the equilibrium density of the whale population, z is a density dependent exponent, and q is the maximum increase in fecundity of which the population is capable at low densities. It is easy to see that when (4.7.2) holds and when the initial conditions are such that

$$N_{-k}, \ldots, N_{-1} \in [0, \infty) \ \text{ and } \ N_0 \in (0, \infty) \qquad (4.7.3)$$

then the corresponding solution $\{N_n\}$ is positive.

Theorem 4.7.1 *Assume that (4.7.2) and (4.7.3) hold and suppose that*

$$0 < qz < 2. \tag{4.7.4}$$

Then the equilibrium K of Eq.(4.7.1) is locally asymptotically stable.

Proof. The linearized equation of Eq.(4.7.1) about the equilibrium K is

$$y_{n+1} - (1 - \mu)y_n + \mu(qz - 1)y_{n-k} = 0, \ n = 0, 1, \ldots. \tag{4.7.5}$$

From Theorem 1.3.7 it follows that Eq.(4.7.5) is asymptotically stable if

$$1 - \mu + \mu|qz - 1| < 1$$

which is equivalent to (4.7.4).   □

The following technical Lemma will be useful in the sequel.

Lemma 4.7.1 *Consider the difference equation*

$$x_{n+1} = x_n\left[1 + q(1 - (\frac{x_n}{K})^z)\right]_+, \ n = 0, 1, \ldots \tag{4.7.6}$$

where $K, q, z \in (0, \infty)$ and

$$q < \frac{(1 + z)^{1+1/z}}{z} - 1. \tag{4.7.7}$$

(a) Set

$$N^* = K\left(\frac{1 + q}{q}\right)^{1/z}. \tag{4.7.8}$$

If $x_0 \in (0, N^*)$, then $x_n \in (0, N^*)$ for all n.

(b) Let $a \leq \min\{K, N^* - K\}$ and assume that

$$q \leq \min\left\{\frac{2}{z}, \ \frac{\frac{2a}{K+a}}{(\frac{K+a}{K})^z - 1}\right\} < \frac{(1+z)^{1+1/z}}{z} - 1. \tag{4.7.9}$$

Then the positive equilibrium K of Eq.(4.7.6) is globally asymptotically stable relative to the interval $(K - a, K + a) \subset (0, N^*)$.

Proof. (a) Consider the function

$$f(x) = x\left[1 + q(1 - (\tfrac{x}{K})^z)\right]_+, \quad \text{for} \ x \in [0, \infty).$$

It is easy to see that $f(0) = 0$, $f(x) > 0$ for $x \in (0, N^*)$, and $f(x) = 0$ for $x \in [N^*, \infty)$. To complete the proof it remains to show that $f(x) \in (0, N^*)$ for $x \in (0, N^*)$. The function f attains a maximum at

$$x_m = K \left(\frac{1+q}{q(1+z)}\right)^{1/z}$$

and

$$f(x_m) = K \left(\frac{1+q}{q(1+z)}\right)^{1/z} \left(\frac{(1+q)z}{1+z}\right).$$

From (4.7.7) it follows that $f(x_m) < N^*$ and the proof of (a) is complete.
(b) We will show that the function v defined by

$$v(x) = |x - K| \quad \text{for} \ x \in (K - a, K + a) \tag{4.7.10}$$

is a Liapunov function for Eq.(4.7.6). Clearly, v is nonnegative and continuous function and so it remains to prove that

$$v(f(x)) < v(x) \quad \text{for} \ x \in (K - a, K + a), \ x \neq K. \tag{4.7.11}$$

Let $x \in (K - a, K)$. Then (4.7.11) is equivalent to

$$x - K < x\left(1 + q\left(1 - \left(\tfrac{x}{K}\right)^z\right)\right) - K < K - x$$

and this is true provided that the function

$$g(x) = 2x + qx\left(1 - \left(\tfrac{x}{K}\right)^z\right) - 2K \tag{4.7.12}$$

is negative for $x \in (0, N^*)$. This follows from the observations that $g(0) = -2K < 0$, $g(K) = 0$, and (by using (4.7.4))

$$g'(x) = 2 + q - q(1 + z)\left(\tfrac{x}{K}\right)^z > 2 + q - q(1 + z) \geq 0.$$

Next, let $x \in (K, K + a)$. Then (4.7.11) is equivalent to

$$K - x < x\left(1 + q\left(1 - \left(\tfrac{x}{K}\right)^z\right)\right) - K < x - K.$$

and this is true provided that $g(x) > 0$ for $x \in (K, a)$. Indeed,

$$g(K) = 0, \quad g(K + a) = 2a + q(K + a)(1 - (\frac{K + a}{K})^z) \geq 0,$$

$$g'(K) = 2 + q - q(1 + z) \geq 0,$$

and the function $g'(x)$ vanishes only at

$$x = \left(\frac{2 + q}{q(1 + z)}\right)^{1/z}.$$

Therefore $g(x) > 0$ for $x \in (K, K + a)$. From this we see that $f((K - a, K + a)) \subset (K - a, K + a)$ and so the result follows from Theorem 1.3.1.   □

Theorem 4.7.2 *Let $a < \min\{K, N^* - K\}$ and assume that (4.7.2) and (4.7.9) hold. Then the positive equilibrium K of Eq.(4.7.1) is globally asymptotically stable relative to the interval $(K - a, K + a) \subset (0, N^*)$.*

Proof. From the proof of Lemma 4.7.1 it follows that there exists a convex Liapunov function v such that (4.7.10) holds. The proof is now a direct consequence of Theorem 2.4.2.   □

Remark 4.7.1 When $a = K$, (4.7.9) reduces to

$$q \leq \min\left\{\frac{2}{z}, \frac{1}{2^z - 1}\right\} < \frac{(1 + z)^{1 + 1/z}}{z} - 1$$

which is the condition for the global asymptotic stability of Eq.(4.7.1) relative to $(0, 2K)$ that was obtained in Fisher [1].

Research Project 4.7.1 *Study the oscillatory behavior of Eq.(4.7.1).*

Research Project 4.7.2 *Obtain conditions for the global asymptotic stability of the equilibrium K of Eq.(4.7.1) relative to the interval $(0, N^*)$.*

4.8 A Semidiscretization of a Delay Logistic Model

Consider the delay logistic equation with several delays

$$\dot{N}(t) = rN(t)(1 - \sum_{i=0}^{m} p_i N(t - \tau_i)), \ t \geq 0 \tag{4.8.1}$$

where

$$r, p_m \in (0, \infty), \quad p_0, \ldots, p_{m-1} \in [0, \infty), \quad \text{and} \quad 0 \leq \tau_0 < \cdots < \tau_m.$$

Let $[\cdot]$ denote the greatest integer function. Then the delay logistic equation with piecewise constant arguments

$$\dot{N}(t) = rN(t)(1 - \sum_{i=0}^{m} p_i N([t - i])), \ t \geq 0 \tag{4.8.2}$$

where

$$r, p_m \in (0, \infty) \quad \text{and} \quad p_0, \ldots, p_{m-1} \in [0, \infty) \tag{4.8.3}$$

may be viewed as a semidiscretization of Eq.(4.8.1). One can see that Eq.(4.8.2) leads to the difference equation

$$N_{n+1} = N_n \exp\{r(1 - \sum_{i=0}^{m} p_i N_{n-i})\}, \ n = 0, 1, \ldots \tag{4.8.4}$$

whose oscillatory and stability properties completely characterize the same properties for Eq.(4.8.2). See Györi and Ladas [2, Section 8.2].

It is easy to see that the unique positive equilibrium $\bar{N}$ of Eq.(4.8.4) is

$$\bar{N} = \frac{1}{P}, \quad \text{where} \quad P = \sum_{i=0}^{m} p_i. \tag{4.8.5}$$

The main result in this section is the following.

Theorem 4.8.1 *Assume that (4.8.3) holds and suppose that*

$$r \leq \frac{P}{mP - (m-1)p_0 + p_m}. \tag{4.8.6}$$

Then $\bar{N}$ is a global attractor of all positive solutions of Eq.(4.8.4).

Proof. We will apply Theorem 2.3.1. The function $G(x, y)$ which is defined by (2.3.2) takes the form

$$G(x, y) = Ay \exp\{-r(p_0 + p_m)y\} \exp\{-rm(P - p_0)x\}$$

where

$$A = \exp\{r(m + 1 - (m - 1)p_0\bar{x} - (P - p_m)\bar{x})\}.$$

Furthermore for $x \le y \le \bar{x}$ we have

$$Ay \exp\{-r(p_0 + p_m)y\} \le \bar{x} \exp\{-r(p_0 + p_m)x\}$$

and for $\bar{x} \le y \le x$ we find

$$Ay \exp\{-r(p_0 + p_m)y\} \ge \bar{x} \exp\{-r(p_0 + p_m)x\}.$$

Consider the function

$$H(x) = \bar{x} \exp(-b(x - \bar{x})) \tag{4.8.7}$$

where

$$b = r(mP - (m - 1)p_0 + p_m). \tag{4.8.8}$$

From the above it follows that

$$G(x, y) \le H(x) \quad \text{for} \quad x \le y \le \bar{x}$$

and

$$G(x, y) \ge H(x) \quad \text{for} \quad \bar{x} \le y \le x$$

and the function F which is defined by (2.3.1) satisfies

$$F(x) \le H(x) \quad \text{for} \quad x \le \bar{x} \quad \text{and} \quad F(x) \ge H(x) \quad \text{for} \quad \bar{x} \le x.$$

According to Lemma 1.6.4, it suffices to show that under the condition (4.8.6), the only positive solution of the equation $H^2(x) = x$ is $x = \bar{x}$. To this end consider the function,

$$h(x) = \ln H^2(x) - \ln x = \ln \bar{x} - \ln x - b\bar{x}(e^{-b(x-\bar{x})} - 1).$$

Then $h(0+) > 0$, $h(\infty) < 0$, $h(\bar{x}) = 0$, and

$$h'(x) = \frac{1}{x}(xb^2\bar{x}e^{-b(x-\bar{x})} - 1).$$

The condition (4.8.6) is equivalent to $b\bar{x} \le 1$, and we have

$$\max_{x \ge 0} \left\{ xb^2\bar{x}e^{-b(x-\bar{x})} \right\} = b\bar{x}e^{b\bar{x}-1} \le 1.$$

Thus $h'(x) \le 0$, with equality if and only if $b\bar{x} = 1$ and $x = \bar{x}$. Hence h is a strictly decreasing function and the proof is complete.   □

4.9 A Discrete Analogue of the Emden-Fowler Equation

Our aim in this section is to investigate the oscillation of solutions of the difference equation

$$\Delta^2 x_{n-1} + p_n x_n^\gamma = 0, \ n = 1, 2, \ldots \tag{4.9.1}$$

where $\Delta^2 x_{n-1} = \Delta(x_n - x_{n-1}) = x_{n+1} - 2x_n + x_{n-1}$,

$$\left.\begin{array}{c} \gamma \text{ is the quotient of odd positive integers} \\[1mm] \text{and } p_n \in [0, \infty) \text{ for } n = 1, 2, \ldots \\[1mm] \text{with } p_n \text{ not eventually equal to zero.} \end{array}\right\} \tag{4.9.2}$$

Eq.(4.9.1) is a discrete analogue of the Emden-Fowler equation

$$y'' + p(t)y^\gamma = 0$$

which appears in astrophysics, relativistic mechanics, nuclear physics, chemical reactions, etc.

The main result in this section is the following theorem which was elegantly established by Hooker and Patula [1].

Theorem 4.9.1 *Assume that (4.9.2) holds. Then the following statements are true:*

(a) Assume that

$$\sum_{n=1}^{\infty} n p_n < \infty. \tag{4.9.3}$$

Then Eq.(4.9.1) has a bounded nonoscillatory solution.

(b) Assume that $\gamma > 1$. Then every solution of Eq.(4.9.1) oscillates if and only if

$$\sum_{n=1}^{\infty} n p_n = \infty. \tag{4.9.4}$$

(c) Assume that $\gamma = 1$ and

$$\sum_{n=1}^{\infty} p_n = \infty. \tag{4.9.5}$$

Then every solution of Eq.(4.9.1) oscillates.

(d) Assume that $0 < \gamma < 1$. Then every solution of Eq.(4.9.1) oscillates if and only if

$$\sum_{n=1}^{\infty} n^\gamma p_n = \infty. \tag{4.9.6}$$

Before we present the proof of the theorem we need the following lemma.

Lemma 4.9.1 *Assume that (4.9.2) holds and that Eq.(4.9.1) has a nonoscillatory solution. Then Eq.(4.9.1) has a solution $\{x_n\}$ such that for some $N \geq 0$,*

$$x_n > 0, \quad \Delta x_n > 0 \quad \text{and} \quad \Delta^2 x_n \leq 0 \quad \text{for} \quad n \geq N. \tag{4.9.7}$$

Proof. As γ is the quotient of odd integers, the opposite of a solution is also a solution and so Eq.(4.9.1) has a solution $\{x_n\}$ which is eventually positive. That is, there exists $N \geq 0$ such that

$$x_n > 0 \quad \text{for} \quad n \geq N.$$

From Eq.(4.9.1) we see that

$$\Delta^2 x_{j-1} = \Delta x_j - \Delta x_{j-1} = -p_j x_j^\gamma \leq 0 \quad \text{for} \quad j \geq 1.$$

Summing up from $j = k+1$ to $j = m$ we find

$$\Delta x_m = \Delta x_k - \sum_{j=k+1}^{m} p_j x_j^\gamma$$

and summing up this from $m = k+1$ to $m = n$ we obtain

$$x_{n+1} - x_{k+1} = (n-k)\Delta x_k - \sum_{j=k+1}^{n} (n-j+1)p_j x_j^\gamma. \tag{4.9.8}$$

We now claim that

$$\Delta x_n > 0 \quad \text{for} \quad n \geq N.$$

Otherwise for some $k \geq N$, $\Delta x_k \leq 0$, and (4.9.8) yields the contradiction that

$$\lim_{n \to \infty} x_n = -\infty.$$

The proof is complete.   □

Proof of the Theorem

(a) Assume that (4.9.3) holds. We must prove that Eq.(4.9.1) has a bounded nonoscillatory solution.

Observe that if $\{x_n\}$ satisfies the equation

$$x_n = 1 - \sum_{i=n+1}^{\infty} (i-n)p_i x_i^{\gamma} \tag{4.9.9}$$

then $\{x_n\}$ is a solution of Eq.(4.9.1). Therefore it suffices to show that Eq.(4.9.9) has a bounded nonoscillatory solution. To this end, choose N so large that

$$\max\left\{ \sum_{i=N}^{\infty} ip_i, 2\gamma \sum_{i=N}^{\infty} ip_i \right\} < \frac{1}{2}.$$

Consider the Banach space ℓ_{∞}^N of all bounded, real sequences $z = \{z_n\}_{n \geq N}$ with the norm defined by

$$\|z\| = \sup_{n \geq N} |z_n|.$$

Set

$$S = \left\{ z \in \ell_{\infty}^N : \frac{1}{2} \leq z_n \leq 1 \ \text{ for } \ n \geq N \right\}.$$

Clearly S is a closed subset of ℓ_{∞}^N and so S is a complete metric space with the distance $d(z', z'') = \|z' - z''\|$. Define the mapping T on S by

$$(Tz)_n = 1 - \sum_{i=n+1}^{\infty} (i-n)p_i z_i^{\gamma} \ \text{ for } \ n \geq N.$$

Note that $z_i^{\gamma} \leq 1$ and so

$$(Tz)_n \geq 1 - \sum_{i=n+1}^{\infty} (i-n)p_i \geq \frac{1}{2} \ \text{ for } \ n \geq N.$$

Also clearly, $(Tz)_n \leq 1$. Thus $T : S \to S$. We now claim that T is a contraction on S. By applying the mean value theorem to the function $f(x) = x^{\gamma}$ we find for $x_1, x_2 \in (1/2, 1)$,

$$|x_1^{\gamma} - x_2^{\gamma}| \leq |f'(\xi)||x_1 - x_2| \ \text{ where } \ \xi \in (\min\{x_1, x_2\}, \max\{x_1, x_2\}).$$

But

$$|f'(\xi)| = |\gamma\xi^{\gamma-1}| \leq \begin{cases} \gamma & \text{if} \quad \gamma \geq 1 \\ 2\gamma & \text{if} \quad 0 < \gamma < 1 \end{cases}$$

and so
$$|x_1^\gamma - x_2^\gamma| \leq 2\gamma|x_1 - x_2| \quad \text{for} \quad x_1, x_2 \in (1/2, 1).$$

Let $z, w \in S$. Then for $n \geq N$,

$$
\begin{aligned}
|(Tz)_n - (Tw)_n| &\leq \sum_{i=n+1}^{\infty} (i-n)p_i|z_i^\gamma - w_i^\gamma| \\
&\leq 2\gamma \sum_{i=n+1}^{\infty} (i-n)p_i|z_i - w_i| \\
&\leq \|z-w\|2\gamma \sum_{i=n+1}^{\infty} (i-n)p_i \leq \frac{1}{2}\|z-w\|.
\end{aligned}
$$

Hence
$$\|Tz - Tw\| \leq \frac{1}{2}\|z - w\|$$

and so T is a contraction on S. The (unique) fixed point of T is the desired bounded, nonoscillatory solution of Eq.(4.9.9).

(b) Assume that (4.9.4) holds. We must prove that every solution of Eq.(4.9.1) oscillates. Otherwise by Lemma 4.9.1, Eq.(4.9.1) has a solution $\{x_n\}$ such that (4.9.7) holds. By multiplying both sides of Eq.(4.9.1) by $nx_n^{-\gamma}$ and then by summing up we obtain

$$\sum_{n=N}^{k-1} nx_n^{-\gamma}\Delta^2 x_{n-1} + \sum_{n=N}^{k-1} np_n = 0 \quad \text{for} \quad k > N.$$

By using the summation-by-parts formula

$$\sum_{n=N}^{k-1} u_n\Delta v_n = u_k v_k - u_N v_N - \sum_{n=N}^{k-1} v_{n+1}\Delta u_n$$

with $u_n = nx_n^{-\gamma}$ and $v_n = \Delta x_{n-1}$ we find

$$kx_k^{-\gamma}\Delta y_{k-1} - Nx_N^{-\gamma}\Delta y_{N-1} - \sum_{n=N}^{k-1}(\Delta x_n)\Delta(nx_n^{-\gamma}) + \sum_{n=N}^{k-1} np_n = 0.$$

Hence
$$\sum_{n=N}^{\infty}(\Delta x_n)\Delta(nx_n^{-\gamma}) = \infty. \tag{4.9.10}$$

Now observe that
$$\Delta(nx_n^{-\gamma}) = x_{n+1}^{-\gamma} + n\Delta(x_n^{-\gamma}) \leq x_{n+1}^{-\gamma}$$

and so

$$\sum_{n=N}^{k} (\Delta x_n)\Delta(nx_n^{-\gamma}) \le \sum_{n=N}^{k} x_{n+1}^{-\gamma}\Delta x_n. \qquad (4.9.11)$$

Set $f(x) = x_n + (\Delta x_n)(x - n)$ for $n \le x \le n+1$ and $n \ge N$. Then f is continuous and increasing for $x \ge N$, and so

$$\begin{aligned}
x_{n+1}^{-\gamma}\Delta x_n &= \int_n^{n+1} x_{n+1}^{-\gamma}\Delta x_n dx \\
&= \int_n^{n+1} f(n+1)^{-\gamma} f'(x) dx \\
&< \int_n^{n+1} f(x)^{-\gamma} f'(x) dx \\
&= \frac{1}{\gamma - 1}[f(n+1)^{1-\gamma} - f(n)^{1-\gamma}].
\end{aligned}$$

By summing up from $n = N$ to $n = k$ we obtain

$$\sum_{n=N}^{k} x_{n+1}^{-\gamma}\Delta x_n \le \frac{1}{\gamma - 1}[f(k+1)^{1-\gamma} - f(N)^{1-\gamma}]. \le \frac{f(N)^{1-\gamma}}{\gamma - 1}$$

which because of (4.9.11) contradicts (4.9.10).

Conversely we must prove that if every solution of Eq.(4.9.1) oscillates and $\gamma > 1$, then (4.9.4) holds. Otherwise (4.9.3) holds and by (a) we obtain the contradiction that Eq.(4.9.1) has a nonoscillatory solution.

(c) Otherwise by Lemma 4.9.1, there exists a solution $\{x_n\}$ such that (4.9.7) holds. In particular $\{x_n\}$ is increasing for $n \ge N$. Hence

$$\Delta x_n - \Delta x_{n-1} + p_n x_N \le 0 \quad \text{for} \quad n \ge N.$$

By summing up from $n = N$ to $n = k$ we have

$$\Delta x_k - \Delta x_{N-1} + x_N \sum_{n=N}^{k} p_n \le 0$$

which, in view of (4.9.5), for k sufficiently large implies $\Delta x_k < 0$. This contradicts (4.9.7) and the proof of (c) is complete.

(d) Assume that (4.9.6) holds. We shall prove that every solution of Eq.(4.9.1) oscillates. Otherwise by Lemma 4.9.1, Eq.(4.9.1) has a solution $\{x_n\}$ such that (4.9.7) holds. Observe that for $n \ge 2N$,

$$x_n = x_N + \sum_{k=N}^{n-1} \Delta x_k \ge x_N + (\Delta x_{n-1})(n - N) \ge \frac{n}{2}(\Delta x_{n-1})$$

and so

$$\frac{x_n}{(\Delta x_{n-1})} \geq \frac{n}{2} \quad \text{for} \quad n \geq 2N. \tag{4.9.12}$$

By dividing both terms of Eq.(4.9.1) by $(\Delta x_{n-1})^\gamma$, and then by applying (4.9.12), and finally by summing up from $n = 2N$ to $n = k$, we obtain

$$\sum_{n=2N}^{k} \frac{\Delta^2 x_{n-1}}{(\Delta x_{n-1})^\gamma} + \frac{1}{2^\gamma} \sum_{n=2N}^{k} n^\gamma p_n \leq 0 \quad \text{for} \quad k \geq 2N.$$

In view of (4.9.6) it follows that

$$\sum_{n=2N}^{k} \frac{\Delta^2 x_{n-1}}{(\Delta x_{n-1})^\gamma} = -\infty. \tag{4.9.13}$$

Now consider again the function $f(x) = x_n + (\Delta x_n)(x - n)$ for $n \leq x \leq n+1$ and $n \geq N$ and set $g(x) = f(x+1) - f(x)$ for $x \geq N$. Then g is continuous, positive, and for $n - 1 < x < n$,

$$g'(x) = f'(x+1) - f(x) = \Delta^2 x_{n-1} \leq 0.$$

Hence g is nonincreasing and so for $n - 1 \leq x \leq n$,

$$g(x) \leq g(n-1) = f(n) - f(n-1) = \Delta x_{n-1}.$$

Then for $n - 1 \leq x \leq n$ we have

$$
\begin{aligned}
\frac{\Delta^2 x_{n-1}}{(\Delta x_{n-1})^\gamma} &= \int_{n-1}^{n} \frac{\Delta^2 x_{n-1}}{(\Delta x_{n-1})^\gamma} dx \\
&\geq \int_{n-1}^{n} \frac{g'(x)}{[g(x)]^\gamma} dx \\
&= \frac{g(n)^{1-\gamma} - g(n-1)^{1-\gamma}}{1 - \gamma}.
\end{aligned}
$$

By summing up from $n = 2N$ to $n = k$ this implies that

$$\sum_{n=2N}^{k} \frac{\Delta^2 x_{n-1}}{(\Delta x_{n-1})^\gamma} \geq \frac{g(k)^{1-\gamma} - g(2N-1)^{1-\gamma}}{1-\gamma} \geq \frac{-g(2N-1)^{1-\gamma}}{1-\gamma}$$

which, as $k \to \infty$, contradicts (4.9.13).

Conversely, we should prove that if every solution of Eq.(4.9.1) oscillates, then (4.9.6) holds. Otherwise

$$\sum_{n=1}^{\infty} n^\gamma p_n < \infty.$$

Now choose N_0 so large that

$$\sum_{n=N_0}^{\infty} n^{\gamma} p_n < \frac{1}{2}$$

and let $\{x_n\}$ be the unique solution of Eq.(4.9.1) with

$$x_{N_0} = 0 \quad \text{and} \quad x_{N_0+1} = 1.$$

One can show by induction that

$$1/2 \leq \Delta x_n \leq 1 \quad \text{for} \quad n \geq N_0$$

and so $\{x_n\}$ is a nonoscillatory solution of Eq.(4.9.1). This contradiction completes the proof of the part (d). The proof of Theorem 4.9.1 is complete. □

Remark 4.9.1 It was shown in Hooker and Patula [1] that when γ is the quotient of odd positive integers and $p_n \leq 0$ for $n = 1, 2, \ldots$ with p_n not eventually equal to zero, then every nontrivial solution of Eq.(4.9.1) is nonoscillatory and eventually monotonic.

Remark 4.9.2 From Eq.(4.9.1) we see that if $x_N = 0$ for some $N \geq 1$, then $x_{N+1} = -x_{N-1}$ and so a nontrivial solution of Eq.(4.9.1) must change sign infinitely often.

Remark 4.9.3 It follows from Theorem 4.9.1(a) that when $\gamma = 1$,

$$\sum_{n=1}^{\infty} np_n = \infty$$

is a necessary condition for all solutions to oscillate. But is it sufficient?

It is interesting to note that when $\gamma \neq 1$, that is when Eq.(4.9.1) is nonlinear, we know necessary and sufficient conditions for oscillation. However, for the linear equation, that is, when $\gamma = 1$, necessary and sufficient conditions for oscillation are not known.

Next, we will consider the special case of Eq.(4.9.1)

$$\Delta^2 x_{n-1} + p x_n^{\gamma} = 0, \; n = 1, 2, \ldots \tag{4.9.14}$$

where $p_n = p > 0$ for $n = 1, 2, \ldots$ and γ is a quotient of two odd integers. The following lemma shows that Eq.(4.9.14) has a periodic solutions with period 2.

Lemma 4.9.2 *Assume $p > 0$ and $\gamma \neq 1$ is a quotient of two odd integers. Set*

$$x_n = (-1)^n (\frac{4}{p})^{1/(\gamma-1)} \quad \text{for} \quad n = 1, 2, \ldots . \tag{4.9.15}$$

Then $\{x_n\}$ is a solution of Eq.(4.9.14) with period 2.

Proof. The assertion of the lemma can be easily verified by a direct substitution of (4.9.15) into Eq.(4.9.14).   □

We pose the following research projects.

Research Project 4.9.1 *Assume $p > 0$ and γ is a quotient of two odd integers. Investigate the existence and stability of periodic solutions of Eq.(4.9.14).*

Research Project 4.9.2 *Assume that (4.9.2) holds and let $\{p_n\}$ be a positive periodic sequence. Investigate the existence and stability of periodic solutions of Eq. (4.9.1).*

4.10 Notes

The results in Section 4.1 are from Kocic and Ladas [3] and Kuruklis and Ladas [1]. See also Jaroma, Kuruklis and Ladas [1] and Kocic, Ladas and Rodrigues [1]. Sections 4.2 and 4.3 are from Grove, Kocic, Ladas and Levins [1,2]. Theorem 4.4.1 is new. The results in Section 4.5 are from Kocic and Ladas [1]. Theorems 4.6.1, 4.6.3 are from Kocic and Ladas [5]. The remaining results in Section 4.6 are from DeVault, Kocic and Ladas [1]. The results in Section 4.7 are from Clark [1] and Fisher [1]. Theorem 4.8.1 is new. For some additional results see Seifert [1, 2]. The results in Section 4.9 are extracted from Hooker and Patula [1]. For some related results see Agarwal [1], Drozdowicz and Popenda [1], Erbe and Zhang [1], Szafranski and Szmanda [1-3], and Szmanda [1,2].

Chapter 5

Periodic Cycles

In this chapter we investigate the behavior of solutions of the difference equation

$$x_{n+1} = \frac{1 + x_n + x_{n-1} + \cdots + x_{n-k+2}}{x_{n-k+1}}, \quad n = 0, 1, \ldots \tag{5.0.1}$$

where k is a nonnegative integer and $x_{-k}, \ldots x_0$ are arbitrary positive numbers.

For $k = 0, 1$, and 2, the special cases of Eq.(5.0.1)

$$x_{n+1} = \frac{1}{x_n}, \quad n = 0, 1, \ldots \tag{5.0.2}$$

$$x_{n+1} = \frac{1 + x_n}{x_{n-1}}, \quad n = 0, 1, \ldots \tag{5.0.3}$$

$$x_{n+1} = \frac{1 + x_n + x_{n-1}}{x_{n-2}}, \quad n = 0, 1, \ldots \tag{5.0.4}$$

have the property that every solution of the equation is periodic with period 2, 5 and 8, respectively. That is, for $k = 0, 1$, and 2 the positive solutions of Eq.(5.0.1) are all periodic with period $3k + 2$. It remains an open question whether this is true for any other value of k. However, computer work indicates that the positive solutions of Eq.(5.0.1) are not all of the same period for any $k \geq 4$ (which we tried).

The 5–cycle Eq.(5.0.3) was discovered by Lyness [1,2] while he was working on a problem in Number Theory. This same equation has also applications in Geometry (see Leech [1]) and in frieze patterns (see Conway and Coxeter [1,2]).

A **frieze pattern** is a collection of horizontal lines of nonnegative numbers with the property that every four adjacent numbers forming a rhombus,

$$\begin{array}{ccc} & q & \\ p & & r \\ & s & \end{array}$$

except for possible borders of zeros and ones, are positive and satisfy the unimodular equation

$$pr - qs = 1.$$

Coxeter [1] has shown that every frieze pattern is periodic. For example, the frieze pattern which is shown below is of period five.

$$
\begin{array}{ccccccccc}
\cdots & 1 & & 1 & & 1 & & 1 & & 1 & \cdots \\
\cdots & & x_1 & & x_3 & & x_5 & & x_2 & & x_4 & \cdots \\
\cdots & x_5 & & x_2 & & x_4 & & x_1 & & x_3 & & \cdots \\
\cdots & & 1 & & 1 & & 1 & & 1 & & 1 & \cdots
\end{array}
$$

Here $x_1 = p$ and $x_2 = q$ are arbitrary positive numbers, and from the definition of frieze patterns, it follows that

$$x_3 = \frac{1+q}{p}, \quad x_4 = \frac{1+p+q}{pq} \quad \text{and} \quad x_5 = \frac{1+p}{q}.$$

Therefore the above pattern is generated by (5.0.3).

Lyness also discovered that the solutions of the 5–cycle (5.0.3) satisfy the invariance

$$(1 + x_n + x_{n+1})\left(1 + \frac{1}{x_n}\right)\left(1 + \frac{1}{x_{n+1}}\right) = Constant, \quad \text{for} \quad n = 0, 1, \dots.$$

Although for $k \geq 4$ the solutions of Eq.(5.0.1) do not seem (from computer experiments) to be all periodic with the same period, Jianshe and Xuli [1] have shown that the above invariance does extend to all values of k.

In Sections 5.1, 5.2 and 5.3 we investigate the behavior of solutions of equations of slightly more general form than Eqs.(5.0.1),(5.0.3) and (5.0.4), respectively.

Research Project 5.0.1 *Is there a value of $k \geq 3$, such that every solution of Eq.(5.0.1) is periodic with the same period?*

Research Project 5.0.2 *Are there values $\alpha, \beta, \gamma \in (0, \infty)$ such that every solution of the rational recursive sequence*

$$x_{n+1} = \frac{1 + \alpha x_n + \beta x_{n-1} + \gamma x_{n-3}}{x_{n-4}}, \quad n = 0, 1, \dots$$

is periodic with the same period?

5.1 An Invariance for $x_{n+1} = \frac{a+x_n+\cdots+x_{n-k+2}}{x_{n-k+1}}$

In this section we establish an invariant relation for the positive solutions of the recursive sequence

$$x_{n+1} = \frac{a + x_n + \cdots + x_{n-k+2}}{x_{n-k+1}}, \quad n = 0, 1, \ldots \tag{5.1.1}$$

where

$$a \in [0, \infty) \quad \text{and} \quad k \in \{1, 2, \ldots\}. \tag{5.1.2}$$

It will follows as a corollary that every positive solution of Eq.(5.1.1) is bounded from below and from above by positive constants.

Whether there exists an invariance for the more general equation

$$x_{n+1} = \frac{a + b_0 x_n + \cdots + b_{k-1} x_{n-k+1}}{x_{n-k}}, \quad n = 0, 1, \ldots \tag{5.1.3}$$

remains an open question at this time.

Theorem 5.1.1 *Assume that (5.1.2) holds and that $x_{-k+1}, \ldots, x_0$ are arbitrary positive numbers. Then the solution $\{x_n\}$ of Eq.(5.1.1) is such that*

$$\left[a + \sum_{j=0}^{k-1} x_{n(k+1)+j} \right] \prod_{j=0}^{k-1} \left[1 + \frac{1}{x_{n(k+1)+j}} \right] = Constant, \quad n = 0, 1, \ldots. \tag{5.1.4}$$

Proof. It follows from Eq.(5.1.1) that for every $n = 0, 1, \ldots$ and $j = 0, 1, \ldots,$

$$x_{n(k+1)+j} = \frac{a + x_{n(k+1)+j-1} + \cdots + x_{n(k+1)+j-k+1}}{x_{n(k+1)+j-k}}$$

and so,

$$1 + x_{n(k+1)+j} = \frac{x_{n(k+1)+j-k} + a + x_{n(k+1)+j-1} + \cdots + x_{n(k+1)+j-k+1}}{x_{n(k+1)+j-k}}$$

$$= \frac{x_{n(k+1)+j-1} + [a + x_{n(k+1)+j-2} + \cdots + x_{n(k+1)+j-k}]}{x_{n(k+1)+j-k}}$$

$$= \frac{x_{n(k+1)+j-1}[1 + x_{(k+1)(n-1)+j}]}{x_{(k+1)(n-1)+j+1}}.$$

Therefore for every $n = 0, 1, \ldots$ and $j = 0, 1, \ldots, k,$

$$\frac{1 + x_{n(k+1)+j}}{x_{n(k+1)+j-1}} = \frac{1 + x_{(k+1)(n-1)+j}}{x_{(k+1)(n-1)+j+1}}. \tag{5.1.5}$$

By multiplying the above equality, term by term, for $j = 0, 1, \ldots, k$ and then by cancelling the common denominators we obtain $P_n = P_{n-1}$ for $n = 0, 1, 2, \ldots$ where we define P_n to be the product

$$P_n = [1 + x_{n(k+1)}] \prod_{j=0}^{k-1} \left[1 + \frac{1}{x_{n(k+1)+j}}\right] [1 + x_{n(k+1)+k}]. \qquad (5.1.6)$$

Finally, observe that the last factor in (5.1.6) is equal to

$$1 + \frac{a + x_{n(k+1)+k-1} + \cdots + x_{n(k+1)+1}}{x_{n(k+1)}}.$$

From the above we see that (5.1.4) is true with the constant in (5.1.4) equal to P_{-1} where

$$P_{-1} = (a + x_{-k+1} + \cdots + x_0)(1 + \frac{1}{x_{-k+1}}) \cdots (1 + \frac{1}{x_0}). \qquad (5.1.7)$$

$\square$

Remark 5.1.1 For $a = 1$ and $k = 2$, the invariance (5.1.4) appears in Lyness [2]. For $a = 1$ and $k \geq 2$, (5.1.4) is due to Jianshe and Xuli [1]. The proof of Theorem 5.1.1 is only a slight modification of the proof in Jianshe and Xuli [1].

Corollary 5.1.1 *Assume that (5.1.2) holds and let $\{x_n\}$ be a positive solution of Eq.(5.1.1). Then there exist positive numbers m and M such that*

$$m \leq x_n \leq M, \; n = 0, 1, \ldots.$$

Finally we present an oscillation result for Eq.(5.1.3).

Theorem 5.1.2 *Assume that*

$$a, b_0, \ldots, b_{k-1} \in (0, \infty) \; \text{with} \; a + \sum_{i=0}^{k-1} b_i > 0 \; \text{and} \; k \in \{1, 2, \ldots\}. \qquad (5.1.8)$$

Then every nontrivial solution of Eq.(5.1.3) is strictly oscillatory about the positive equilibrium $\bar{x}$ of the equation. Furthermore, every semicycle of a nontrivial solution contains no more than $2k + 1$ terms.

Proof. Let $\{x_n\}$ be a nontrivial solution of Eq.(5.1.3) and assume for the sake of contradiction that $\{x_n\}$ is not strictly oscillatory about the positive equilibrium $\bar{x}$ of the equation. Then there exists $n_0 \geq k$ such that either

$$x_n \geq \bar{x} \quad \text{for} \quad n \geq n_0 - k - 1 \tag{5.1.9}$$

or

$$x_n \leq \bar{x} \quad \text{for} \quad n \geq n_0 - k - 1. \tag{5.1.10}$$

We will assume that (5.1.9) holds. The case where (5.1.10) holds is similar and will be omitted.

Let A be the set of points

$$\{x_{n_0-k}, \ldots, x_{n_0}, \ldots, x_{n_0+k}\}$$

and let j be the smallest integer in the set $\{n_0 - k, \ldots, n_0 + k\}$ with the property that $x_j = \max A$. Now if $j < n_0$, then

$$x_{j+k+1} = \frac{a + \sum_{i=0}^{k-1} b_i x_{j+k-i}}{x_j} < \frac{a}{\bar{x}} + \sum_{i=0}^{k-1} b_i = \bar{x}$$

while if $j \geq n_0$, then

$$x_{j-k-1} = \frac{a + \sum_{i=0}^{k-1} b_i x_{j-i-1}}{x_j} < \frac{a}{\bar{x}} + \sum_{i=0}^{k-1} b_i = \bar{x}.$$

In either case we are led to a contradiction and so the proof that every nontrivial solution of Eq.(5.1.3) is strictly oscillatory is complete. The above analysis also shows that no semicycle may contain more than $2k + 1$ terms. $\square$

Research Project 5.1.1 *Obtain, if possible, an invariance for the rational recursive sequence*

$$x_{n+1} = \frac{1 + \alpha x_n + \beta x_{n-1}}{x_{n-2}}, \quad n = 0, 1, \ldots$$

where $\alpha, \beta \in (0, \infty)$.

5.2 The Five–Cycle: $x_{n+2} = \frac{1+x_{n+1}}{x_n}$

In this section we study the recursive sequence

$$x_{n+1} = \frac{a + bx_n}{x_{n-1}}, \quad n = 0, 1, \dots \tag{5.2.1}$$

where

$$a, b \in [0, \infty) \quad \text{with} \quad a + b > 0$$

and where x_{-1} and x_0 are arbitrary positive numbers.

When $a = 0$ and $b \in (0, \infty)$ every solution of Eq.(5.2.1) is periodic with period six.

When $a \in (0, \infty)$ and $b = 0$ every solution of Eq.(5.2.1) is periodic with period four.

So in the sequel we will assume that

$$a, b \in (0, \infty). \tag{5.2.2}$$

The change of variables $x_n = by_n$ reduces Eq.(5.2.1) to

$$x_{n+1} = \frac{\alpha + x_n}{x_{n-1}}, \quad n = 0, 1, \dots \tag{5.2.3}$$

where $\alpha = a/b^2$. The following result about Eq.(5.2.1) is an elementary consequence of Corollary 5.1.1 and Theorem 5.1.2.

Corollary 5.2.1 *Assume that (5.2.2) holds. Then the following statements are true:*

(a) Every positive solution of Eq.(5.2.1) is bounded away from zero and infinity by positive constants.

(b) Every positive nontrivial solution of Eq.(5.2.1) is strictly oscillatory about the positive equilibrium $\bar{x}$ of Eq.(5.2.1). Furthermore, every semicycle of a nontrivial solution contains at most three terms.

Part (b) of Corollary 5.2.1 may be strengthened as follows:

Theorem 5.2.1 *Assume that (5.2.2) holds. Then the following statements about the positive solutions of Eq.(5.2.1) are true:*

(a) The absolute extreme in a semicycle occurs in the first or in the second term.

(b) Every nontrivial semicycle, after the first one, contains at least two and at most three terms.

Proof. We give the proofs for positive semicycles. The proofs for negative semicycles are similar and will be omitted.

(a) Suppose that x_n, x_{n+1}, and x_{n+2} are terms in a nontrivial positive semicycle. It suffices to show that

$$x_{n+2} < x_{n+1}. \tag{5.2.4}$$

As the semicycle is nontrivial, $\bar{x} \le x_n$ and $x_{n+1} > \bar{x}$. Then

$$\frac{x_{n+2}}{x_{n+1}} = \frac{1}{x_n}\left(\frac{a}{x_{n+1}} + b\right) < \frac{1}{\bar{x}}\left(\frac{a}{\bar{x}} + b\right) = 1$$

and so (5.2.4) is true.

(b) Assume for the sake of contradiction that x_n is the only term in a nontrivial positive semicycle, after the first semicycle. Then

$$x_{n-1} < \bar{x}, \ x_n \ge \bar{x} \quad \text{and} \quad x_{n+1} < \bar{x}$$

and so

$$x_{n+1} = \frac{a + bx_n}{x_{n-1}} > \frac{a + b\bar{x}}{\bar{x}} = \bar{x}$$

which is a contradiction. By Corollary 5.2.1(b), a nontrivial semicycle, after the first semicycle, contains at most three terms. The proof is complete. $\square$

Next we establish a Lemma which implies, among other things, the periodicity result of Lyness which we mentioned in the introduction.

Lemma 5.2.1 *Assume that $\alpha \ge 0$ and let $\{x_n\}$ be a positive solution of Eq.(5.2.3). Then for $n = 0, 1, \ldots$, the following statements are true:*

$$y_{n+5} - y_n = (1 - \alpha)\left(\frac{1}{y_{n+2}} - \frac{1}{y_{n+3}}\right); \tag{5.2.5}$$

$$y_{n+6} - y_n = \alpha\left(1 + \frac{1}{y_{n+3}}\right)\left(\frac{1}{y_{n+4}} - \frac{1}{y_{n+2}}\right). \tag{5.2.6}$$

Proof. We will prove (5.2.5). The proof of (5.2.6) is similar and will be omitted. To this end, from Eq.(5.2.3) we find

$$y_n = \frac{\alpha + y_{n+1}}{y_{n+2}} = \frac{1}{y_{n+2}}\left(\alpha + \frac{\alpha + y_{n+2}}{y_{n+3}}\right) = \frac{\alpha y_{n+3} + \alpha + y_{n+2}}{y_{n+2}y_{n+3}}$$

and

$$y_{n+5} = \frac{\alpha + y_{n+4}}{y_{n+3}} = \frac{1}{y_{n+3}}\left(\alpha + \frac{\alpha + y_{n+3}}{y_{n+2}}\right) = \frac{\alpha y_{n+2} + \alpha + y_{n+3}}{y_{n+2}y_{n+3}}.$$

Eq.(5.2.5) now follows by subtracting and simplifying. $\square$

Corollary 5.2.2 *Assume that* $\alpha \geq 0$. *Then the following statements are true:*

(a) Every positive solution of Eq.(5.2.3) is periodic with period five if and only if

$$\alpha = 1.$$

(b) Every positive solution of Eq.(5.2.3) is periodic with period six if and only if

$$\alpha = 0.$$

Remark 5.2.1 It is interesting to note that the reciprocal of a solution of the equation

$$x_{n+1} = \frac{x_n}{x_{n-1}}, \quad n = 0, 1, \ldots \tag{5.2.7}$$

is also a solution of the same equation. This can be shown directly or by making the also interesting observation that the change of variable $x_n = e^{y_n}$ reduces Eq.(5.2.7) to the linear equation

$$y_{n+1} - y_n + y_{n-1} = 0, \quad n = 0, 1, \ldots. \tag{5.2.8}$$

The periodic characters of Eqs.(5.2.7) can be also deduced from the fact that the characteristic roots of Eq.(5.2.8) are the 6th roots of unity.

Eq.(5.2.3) has a unique positive equilibrium $\bar{y}$. The equilibrium $\bar{y}$ satisfies the equation

$$\bar{y} = \frac{\alpha + \bar{y}}{\bar{y}}$$

and is given by $\bar{y} = (1 + \sqrt{1 + 4\alpha})/2$.

The linearized equation of Eq.(5.2.3) about the positive equilibrium $\bar{y}$ is

$$z_{n+1} - \frac{1}{\bar{y}} z_n + z_{n-1} = 0$$

with characteristic roots

$$\lambda_1 = \frac{1 + i\sqrt{4\bar{y} - 1}}{2\bar{y}} \quad \text{and} \quad \lambda_2 = \frac{1 - i\sqrt{4\bar{y} - 1}}{2\bar{y}}.$$

For $\alpha = 1$, $\lambda_1 = \cos\frac{2\pi}{5} + i\sin\frac{2\pi}{5}$ and $\lambda_2 = \cos\frac{2\pi}{5} - i\sin\frac{2\pi}{5}$ which are two 5th roots of unity.

Thus although linearized stability analysis provides no conclusion about the stability of Eq.(5.2.3), it provides a lead in discovering its periodic nature.

Research Project 5.2.1 *Extend the results of this section to equations where some of the coefficients are negative.*

Research Project 5.2.2 *Extend the results of this section to equations with complex coefficients.*

Research Project 5.2.3 *Assume that the sequences $\{a_n\}$ and $\{b_n\}$ are periodic. Investigate the behavior of solutions of the rational recursive sequence*

$$x_{n+1} = \frac{a_n + b_n x_n}{x_{n-1}}, \ n = 0, 1, \ldots.$$

5.3 The Eight–Cycle: $x_{n+3} = \frac{1+x_{n+2}+x_{n+1}}{x_n}$

In this section we study the recursive sequence

$$x_{n+1} = \frac{a + bx_n + cx_{n-1}}{x_{n-2}}, \ n = 0, 1, \ldots \tag{5.3.1}$$

where

$$a, b, c \in [0, \infty) \ \text{with} \ a + b + c > 0 \tag{5.3.2}$$

and where x_{-2}, x_{-1} and x_0 are arbitrary positive numbers. Eq.(5.3.1) has a unique positive equilibrium $\bar{x}$ which is the positive root of the equation

$$\bar{x} = \frac{a}{\bar{x}} + b + c.$$

The linearized equation of Eq.(5.3.1) about the positive equilibrium $\bar{x}$ is

$$y_{n+1} - \frac{b}{\bar{x}}y_n - \frac{c}{\bar{x}}y_{n-1} + y_{n-2} = 0. \tag{5.3.3}$$

According to Theorem 1.4.2, if all solutions of Eq.(5.3.1) are periodic with the same period p, then it is necessary (but not sufficient) that all solutions of Eq.(5.3.3) be periodic with same period p. The characteristic equation of Eq.(5.3.3) is

$$\lambda^3 - \frac{b}{\bar{x}}\lambda^2 - \frac{c}{\bar{x}}\lambda + 1 = 0 \tag{5.3.4}$$

which has a negative root. If all solutions of Eq.(5.3.3) are periodic with period p, then all roots of Eq.(5.3.4) must be p^{th} roots of unity. In particular, the negative root of Eq.(5.3.4) must be equal to -1. This implies that $b = c$ is a necessary condition for all solutions of Eq.(5.3.1) to be periodic.

When $b = c = 0$, every positive solution of the resulting equation

$$x_{n+1} = \frac{a}{x_{n-2}}, \ n = 0, 1, \ldots \tag{5.3.5}$$

is periodic with period six. Note that the change of variables $x_n = \sqrt{a}e^{y_n}$ reduces Eq.(5.3.5) to the linear equation $y_{n+1} + y_{n-2} = 0$ whose characteristic roots are 6th roots of unity. From this the periodicity of Eq.(5.3.5) also follows, together with the additional fact that when $a = 1$, the reciprocal of a solution of Eq.(5.3.5) is again a solution.

When $b = c \neq 0$, the change of variables $x_n = by_n$ reduces Eq.(5.3.1) to the equation

$$y_{n+1} = \frac{\alpha + y_n + y_{n-1}}{y_{n-2}}, \ n = 0, 1, \ldots \tag{5.3.6}$$

where $\alpha = a/b^2$.

The three roots of the characteristic equation (5.3.4) of Eq.(5.3.3) have product -1. Also Eq.(5.3.4) has a negative root. It follows that a necessary condition for the stability of the equilibrium solution of Eq.(5.3.1) is that -1 be a root of Eq.(5.3.4) or, equivalently, that

$$b = c.$$

If $b \neq c$, Eq.(5.3.4) has a root λ with $|\lambda| > 1$ and so Eq.(5.3.1) is unstable.

From the above discussion and from Corollary 5.1.1 and Theorem 5.1.1 it follows that the following result is true.

Theorem 5.3.1 *Assume that (5.3.2) holds. Then the following statements are true:*

(a) *If $b \neq c$, the positive equilibrium $\bar{x}$ of Eq.(5.3.1) is unstable.*

(b) *If $b = c$, every positive solution of Eq.(5.3.1) is bounded away from zero and infinity by positive constants.*

(c) *Every positive nontrivial solution of Eq.(5.3.1) is strictly oscillatory about the positive equilibrium $\bar{x}$. Furthermore, every semicycle of a nontrivial solution contains at most five terms.*

Remark 5.3.1 Computer observations show that after the first semicycle, every semicycle of a nontrivial solution of Eq.(5.3.1) may contain one, three, four or five terms but never exactly two terms. We will give a proof of this observation for a positive semicycle. The proof for a negative semicycle is similar and will be omitted. To this end, assume that x_{n-1} and x_n are the first two terms in a positive semicycle. Then $x_{n-2} < \bar{x}$ and

$$x_{n+1} = \frac{a + bx_n + cx_{n-1}}{x_{n-2}} > \frac{a + b\bar{x} + c\bar{x}}{\bar{x}} = \bar{x}$$

which proves our claim.

The linearized equation of Eq.(5.3.6) about the positive equilibrium $\bar{y} = 1 + \sqrt{1+\alpha}$ is

$$z_{n+1} - \frac{1}{\bar{y}}z_n - \frac{1}{\bar{y}}z_{n-1} + z_{n-2} = 0.$$

The three characteristic roots associated with this equation are

$$\lambda_1 = -1, \quad \lambda_2 = \frac{1}{2}\left(1 + \frac{1}{\bar{y}} + [(1 + \frac{1}{\bar{y}})^2 - 4]^{1/2}\right)$$

$$\text{and} \quad \lambda_3 = \frac{1}{2}\left(1 + \frac{1}{\bar{y}} - [(1 + \frac{1}{\bar{y}})^2 - 4]^{1/2}\right)$$

and they all lie on the unit disk. For $\alpha = 1$ one can see that these three roots are 8th roots of the unity. This suggests that for $\alpha = 1$, all solutions of Eq.(5.3.6) may be periodic with period eight. In fact the following result is true.

Theorem 5.3.2 *Assume that $a, b \in [0, \infty)$ with $a + b > 0$. Then every positive solution of the recursive sequence*

$$x_{n+1} = \frac{a + bx_n + bx_{n-1}}{x_{n-2}}, \quad n = 0, 1, \ldots \tag{5.3.7}$$

is periodic with period 8 if and only if $a = b^2$.

Proof. It follows from (5.3.7) that

$$x_n = \frac{a + bx_{n+1} + bx_{n+2}}{x_{n+3}}$$

$$= \frac{1}{x_{n+3}}\left(a + b\frac{a + bx_{n+2} + bx_{n+3}}{x_{n+4}} + b\frac{a + bx_{n+3} + bx_{n+4}}{x_{n+5}}\right).$$

Similarly,

$$x_{n+8} = \frac{a + bx_{n+7} + bx_{n+6}}{x_{n+5}}$$

$$= \frac{1}{x_{n+5}}\left(a + b\frac{a + bx_{n+6} + bx_{n+5}}{x_{n+4}} + b\frac{a + bx_{n+5} + bx_{n+4}}{x_{n+3}}\right).$$

Therefore,

$$x_{n+8} - x_n = (a - b^2)\left(1 + \frac{b}{x_{n+4}}\right)\left(\frac{1}{x_{n+5}} - \frac{1}{x_{n+3}}\right)$$

from which the result follows.   □

Research Project 5.3.1 *Extend the results of this section to equations where some of the coefficients are negative.*

Research Project 5.3.2 *Extend the results of this section to equations with complex coefficients.*

5.4   Symmetric Periodic Sequences

Consider the rational recursive sequence

$$x_{n+1} = \frac{a + \sum_{i=0}^{k-1} b_i x_{n-i}}{x_{n-1}}, \quad n = 0, \pm1, \pm2, \ldots \tag{5.4.1}$$

where

$$a \in (0, \infty) \quad \text{and} \quad b_0, \ldots b_{k-1} \in [0, \infty). \tag{5.4.2}$$

Our aim in this section is to show that under appropriate hypotheses, when the linearized equation

$$Ey_{n+1} + Ey_{n-k} = \sum_{i=0}^{k-1} b_i y_{n-i}, \quad n = 0, \pm1, \pm2, \ldots \tag{5.4.3}$$

about the positive equilibrium E of Eq.(5.4.1) has a periodic solution with minimal period $2(k+1)$, then Eq.(5.4.1) also has a periodic solution of minimal period $2(k+1)$.

A finite sequence of real numbers $\{c_l, c_{l+1}, \ldots, c_m\}$ is called **symmetric** if

$$c_i = c_{l+m-i} \quad \text{for} \quad i = l, l+1, \ldots, m.$$

Throughout this section we will assume without further mention that the coefficients $\{b_0, \ldots, b_k\}$ form a symmetric sequence of nonnegative numbers; that is

$$0 \leq b_i = b_{k-1-i} \quad \text{for} \quad i = 0, \ldots, k-1 \tag{5.4.4}$$

and that the initial conditions for a solution of Eq.(5.4.1) are of the form

$$x_n = \varphi_n \quad \text{for} \quad n = 1, \ldots, k+1 \tag{5.4.5}$$

where the numbers φ_n are positive and the sequence $\{\varphi_0, \ldots, \varphi_{k+1}\}$ is symmetric; that is

$$0 < \varphi_i = \varphi_{k+2-i} \quad \text{for } i = 0, \ldots, k+1. \tag{5.4.6}$$

One can now show that with such initial conditions given, Eq.(5.4.1) has a unique solution $\{x_n\}_{n=-\infty}^{\infty}$ which is positive and symmetric in the sense that

$$0 < x_n = x_{k+2-n} \quad \text{for } n = 0, \pm 1 \ldots. \tag{5.4.7}$$

A sequence $\{x_n\}_{n=-\infty}^{\infty}$ is called **periodic of period p** if

$$x_{n+p} = x_n \quad \text{for } n = 0, \pm 1 \ldots. \tag{5.4.8}$$

The least positive number p for which (5.4.8) holds is called the **minimal period** of the sequence.

Eq.(5.4.1) has a unique positive equilibrium E. The equilibrium E satisfies the quadratic equation

$$E^2 - (\sum_{i=0}^{k-1} b_i)E - a = 0$$

and is given by

$$E = \frac{(\sum_{i=0}^{k-1} b_i) + \sqrt{(\sum_{i=0}^{k-1} b_i) + 4a}}{2}. \tag{5.4.9}$$

Eq.(5.4.3) is the linearized equation of Eq.(5.4.1) about E. The main result in this section is the following:

Theorem 5.4.1 *Assume k is odd. Suppose that the linearized equation (5.4.3) has a periodic solution with (minimal) period $2(k+1)$. Then Eq.(5.4.1) has infinitely many symmetric periodic solutions, each with (minimal) period $2(k + 1)$, and arbitrarily near the equilibrium E.*

Proof. The proof is long and will be accomplished in a series of lemmas.

First we will establish a system of algebraic equations which yields symmetric periodic solution of Eq.(5.4.1). Suppose $k = 2m - 1$ is an odd number. If $\{x_n\}_{n=-\infty}^{\infty}$ is a symmetric periodic solution of Eq.(5.4.1) with period $2(k+1) = 4m$, then

$$x_n = x_{2m+1-n} \quad \text{and} \quad x_{n+4m} = x_n \quad \text{for all } n. \tag{5.4.10}$$

Let

$$D = \begin{bmatrix} 0 & \cdots & 0 & 0 & 1 \\ 0 & \cdots & 0 & 1 & 0 \\ 0 & \cdots & 1 & 0 & 0 \\ & \vdots & & & \\ 1 & \cdots & 0 & 0 & 0 \end{bmatrix} \tag{5.4.11}$$

be the $m \times m$-antidiagonal matrix. Define

$$X_i = (x_{im+1}, x_{im+2}, \ldots, x_{im+m})^T \in \mathbf{R}^m \quad \text{for all } i, \tag{5.4.12}$$

and set

$$W = X_2 - X_0 = (x_{2m+1} - x_1, \ldots, x_{3m} - x_m)^T \in \mathbf{R}^m. \tag{5.4.13}$$

Then (5.4.10) yields

$$X_1 = DX_0, \quad X_2 = X_0 + W, \quad X_3 = DX_2 = D(X_0 + W), \tag{5.4.14}$$

and

$$X_{4+i} = X_i \quad \text{for} \quad i = 0, 1, 2, \ldots,$$

where the equality $X_3 = DX_2$ is because $x_{3m+n} = x_{2m+1-3m-n} = x_{2m+m+1-n}$ for $n = 1, 2, \ldots, m$. On the other hand, if X_i satisfies (5.4.14) for some $W \in \mathbf{R}^m$ and $\{x_n\}_{n=-\infty}^{\infty}$ is determined by (5.4.12), then it is easy to verify that $\{x_n\}$ satisfies (5.4.10). Now let

$$B_0 = \begin{bmatrix} 0 & b_0 & b_1 & \ldots & b_{m-3} & b_{m-2} \\ 0 & 0 & b_0 & \ldots & b_{m-4} & b_{m-3} \\ 0 & 0 & 0 & \ldots & b_{m-5} & b_{m-4} \\ & & \ldots & & \ldots & \\ 0 & 0 & 0 & \ldots & 0 & b_0 \\ 0 & 0 & 0 & \ldots & 0 & 0 \end{bmatrix},$$

$$B_2 = \begin{bmatrix} 0 & 0 & 0 & \ldots & 0 & 0 & 0 \\ b_{2m-2} & 0 & 0 & \ldots & 0 & 0 & 0 \\ b_{2m-3} & b_{2m-2} & 0 & \ldots & 0 & 0 & 0 \\ & & \ldots & & & \ldots & \\ b_{m+1} & b_{m+2} & b_{m+3} & \ldots & b_{2m-2} & 0 & 0 \\ b_m & b_{m+1} & b_{m+2} & \ldots & b_{2m-3} & b_{2m-2} & 0 \end{bmatrix}$$

and

$$B_1 = \begin{bmatrix} b_{m-1} & b_m & b_{m+1} & \ldots & b_{2m-3} & b_{2m-2} \\ b_{m-2} & b_{m-1} & b_m & \ldots & b_{2m-4} & b_{2m-3} \\ b_{m-3} & b_{m-2} & b_{m-1} & \ldots & b_{2m-5} & b_{2m-4} \\ & & \ldots & & \ldots & \\ b_1 & b_2 & b_3 & \ldots & b_{m-1} & b_m \\ b_0 & b_1 & b_2 & \ldots & b_{m-2} & b_{m-1} \end{bmatrix}$$

and let $A = (a, a, \ldots, a)^T \in \mathbf{R}^m$. Clearly,

$$DB_0 D = B_2, \quad DB_1 D = B_1 \quad \text{and} \quad DB_2 D = B_0 \tag{5.4.15}$$

or equivalently,

$$DB_0 = B_2 D, \quad DB_1 = B_1 D \quad \text{and} \quad DB_2 = B_0 D.$$

For any $U = (u_1, u_2, \ldots, u_m)^T$ and $V = (v_1, v_2, \ldots, v_m)^T$ in $\mathbf{R}^m$, define

$$U * V = (u_1 v_1, u_2 v_2, \ldots, u_m v_m)^T \in \mathbf{R}^m. \tag{5.4.16}$$

With this notation, Eq.(5.4.1) can be rewritten in the form

$$X_{i+2} * X_i = A + B_0 X_i + B_i X_{i+1} + B_2 X_{i+2}, \; i = 0, 1, \ldots. \tag{5.4.17}$$

For $i = 0$ and 1 Eq.(5.4.17) becomes

$$\left.\begin{aligned}
(X_0 + W) * X_0 &= A + (B_0 + B_1 D + B_2)X_0 + B_2 W \\
(D(X_0 + W)) * (DX_0) &= A + (B_0 DX_0 + B_1(X_0 + W) \\
&\quad + B_2 D(X_0 + W))
\end{aligned}\right\} \tag{5.4.18}$$

where the relations in Eq.(5.4.14) were used. Notice that $D^2 = I$ (identity matrix), $DA = A$, and $(DU) * (DV) = D(U * V)$ for any U, V in $\mathbf{R}^m$. By multiplying by D and by using Eq.(5.4.15), the second equation in Eq.(5.4.18) becomes

$$(X_0 + W) * X_0 = A + (B_2 + B_1 D + B_0)X_0 + B_1 DW + B_0 W.$$

By subtracting this equation from the first equation in Eq.(5.4.18), Eq.(5.4.18) is equivalent to the following equations:

$$\left.\begin{aligned}
(X_0 + W) * X_0 &= A + (B_0 + B_1 D + B_2)X_0 + B_2 W \\
\tilde{B}_1 W &= 0,
\end{aligned}\right\} \tag{5.4.19}$$

where $\tilde{B}_1 = B_0 + B_1 D - B_2$. Therefore a symmetric periodic sequence $\{x_n\}_{n=-\infty}^{\infty}$ with period $4m$ (that is, satisfying Eq.(5.4.10)) is a solution of Eq.(5.4.1) if and only if Eq.(5.4.17) is satisfied for $i = 0, 1, 2, 3$. When $i = 2, 3$, Eq.(5.4.17) becomes

$$\left.\begin{aligned}
X_4 * X_2 &= A + B_0 X_2 + B_1 X_3 + B_2 X_4 \\
X_5 * X_3 &= A + B_0 X_3 + B_1 X_4 + B_2 X_5.
\end{aligned}\right\} \tag{5.4.20}$$

Since

$$X_4 * X_2 = X_0 * X_2 = X_2 * X_0, \quad X_5 * X_3 = X_3 * X_1,$$

$$B_0 X_2 + B_1 X_3 + B_2 X_4 = B_0(X_0 + W) + B_1 D(X_0 + W) + B_2 X_0$$

$$= (B_0 + B_1 D + B_2)X_0 + B_0 W + B_1 DW$$

and

$$B_0 X_3 + B_1 X_4 + BX_5 = B_0 D(X_0 + W) + B_1 X_0 + B_2 DX_0$$

$$= B_0 DX_0 + B_1 X_0 + B_2 DX_0 + B_0 DW,$$

we see that Eq.(5.4.20) is also equivalent to Eq.(5.4.19). In summary, we have established the following lemma.

Lemma 5.4.1 *Suppose $k = 2m - 1$ is an odd number. If $\{x_n\}_{n=-\infty}^{\infty}$ is a symmetric periodic solution of Eq.(5.4.1) with period $4m$, then (X_0, W) as given by Eq.(5.4.12) and Eq.(5.4.13) is a solution of Eq.(5.4.19). On the other hand, if $(X_0, W) \in \mathbf{R}^m \times \mathbf{R}^m$ is a solution of Eq.(5.4.19) and X_i for $i = 1, 2, \ldots$ and $\{x_n\}_{n=-\infty}^{\infty}$ are given by Eq.(5.4.14) and Eq.(5.4.12) respectively, then $\{x_n\}_{n=-\infty}^{\infty}$ is a symmetric periodic solution of Eq.(5.4.1) with period $4m$.*

In a similar way, consider Eq.(5.4.3), and suppose that $\{y_n\}_{n=-\infty}^{\infty}$ is a symmetric periodic solution of Eq.(5.4.3) with period $2(k + 1) = 4m$. Define

$$Y_i = (y_{im+1}, y_{im+2}, \ldots, y_{im+m})^T \in \mathbf{R}^m, i = 0, 1, \ldots \tag{5.4.21}$$

and

$$V = Y_2 - Y_0. \tag{5.4.22}$$

Then Eq.(5.4.14) becomes

$$\left. \begin{array}{c} Y_1 = DY_0, \quad Y_2 = Y_0 + V, \quad Y_3 = D(Y_0 + V), \\ \\ \text{and} \ \ Y_{4+i} = Y_i \ \ \text{for all } i. \end{array} \right\} \tag{5.4.23}$$

As in the above discussion with Eq.(5.4.1), Eq.(5.4.3) is reduced to the following equations:

$$\left. \begin{array}{rcl} E(Y_0 + V) + EY_0 & = & (B_0 + B_1 D + B_2)Y_0 + B_2 V \\ \\ \tilde{B}_1 V & = & 0. \end{array} \right\}$$

That is,

$$\left. \begin{array}{rcl} (2EI - (B_0 + B_1 D + B_2))Y_0 & = & (B_2 - EI)V \\ \\ \tilde{B}_1 V & = & 0, \end{array} \right\} \tag{5.4.24}$$

where $\tilde{B}_1 = B_0 + B_1 D - B_2$.

Lemma 5.4.2 *Suppose $k = 2m - 1$ is an odd number. If $\{y_n\}_{n=-\infty}^{\infty}$ is a symmetric periodic solution of Eq.(5.4.3) with period $4m$ then (Y_0, V) as defined by Eq.(5.4.21) and Eq.(5.4.22) is a solution of Eq.(5.4.24). On the other hand, if $(Y_0, V) \in \mathbf{R}^m \times \mathbf{R}^m$ is a solution of Eq.(5.4.24) and Y_i for $i = 1, 2, \ldots$ and $\{y_n\}_{n=-\infty}^{\infty}$ are defined by Eq.(5.4.23) and Eq.(5.4.21) respectively, then $\{y_n\}_{n=-\infty}^{\infty}$ is a symmetric periodic solution of Eq.(5.4.3) with period $4m$.*

In view of Lemma 5.4.1, finding a symmetric periodic solution of Eq.(5.4.1) with period $2(k+1) = 4m$ is equivalent to finding a solution of Eq.(5.4.19). Since $x_n \equiv E$ is an equilibrium of Eq.(5.4.1), if $\hat{X}_0 = (E, E, \ldots, E)^T \in \mathbf{R}^m$, then $(X_0, W) = (\hat{X}_0, 0)$ is a solution of Eq.(5.4.19). It is clear that the linearization of Eq.(5.4.19) for (X_0, W) around $(\hat{X}_0, 0)$ is just Eq.(5.4.24). Since Eq.(5.4.24) is equivalent to Eq.(5.4.3), by Lemma 5.4.2 we have the following result about the symmetric periodic solutions of Eq.(5.4.3).

Lemma 5.4.3 *Suppose $a > 0$ and $b_i \geq 0$ for $i = 0, 1, 2, \ldots, k-1$. Then the following statements are true:*

(a) Eq.(5.4.3) has a non-trivial periodic solution with period p if and only if for some integer q, $\lambda = e^{2q\pi i/p}$ $(i^2 = -1)$ is a solution of the equation

$$E(\lambda^{k+1} + 1) = \sum_{j=0}^{k-1} b_j \lambda^{k-j}. \tag{5.4.25}$$

(b) Eq.(5.4.3) has no non-trivial periodic solution with period $(k + 1)$.

(c) If $k = 2m - 1$ is odd and Eq.(5.4.3) has a non-trivial periodic solution with (minimal) period $4m$, then Eq.(5.4.3) has a non-trivial symmetric periodic solution with (minimal) period $4m$.

Proof. (a) is obviously true because Eq.(5.4.25) is the characteristic equation of Eq.(5.4.3).

(b) Since

$$\sum_{j=0}^{k-1} b_j \lambda^{k-j} \leq \sum_{j=0}^{k-1} b_j \quad \text{for any} \lambda = 1$$

and

$$E = \frac{\sum_{j=0}^{k-1} b_j + \sqrt{\left(\sum_{j=0}^{k-1} b_j\right)^2 + 4a}}{2} > \sum_{j=0}^{k-1} b_j,$$

there is no solution of Eq.(5.4.25) satisfying $\lambda^{k+1} - 1 = 0$. Therefore it follows from (a) that there is no non-trivial periodic solution of Eq.(5.4.3) with period $(k+1)$.

(c) Finally, if Eq.(5.4.3) has a non-trivial periodic solution of period $2(k+1) = 4m$, then it follows from (a) that there exists a solution $\lambda = e^{2q\pi i/4m}$ of Eq.(5.4.25) for some integer q. According to (b), the integer q is odd. Since $\lambda = e^{-2q\pi i/4m}$ is also a solution of Eq.(5.4.25), $\{c\sin(\frac{nq\pi}{2m} + \theta)\}_{n=-\infty}^{\infty}$ is a periodic solution of Eq.(5.4.3) for any fixed $c, \theta \in \mathbf{R}$. In particular, if $c = 1$ and $\theta = -\frac{q\pi}{2}$, then $\{\sin\frac{(2n-1)q\pi}{4m}\}_{n=-\infty}^{\infty}$ is a symmetric periodic solution of Eq.(5.4.3) with period $4m$. If the non-trivial periodic solution is of minimal period $4m$, then q is relatively prime to $4m$. Therefore the symmetric periodic solution is of minimal period $4m$ also. The proof is complete. $\square$

Now we are ready to establish Theorem 5.4.1. Let $k = 2m - 1$ and suppose that Eq.(5.4.3) has a non-trivial periodic solution $\{y_n\}_{n=-\infty}^{\infty}$ with period $4m$. According to (c) of Lemma 5.4.3, we may assume that $\{y_n\}_{n=-\infty}^{\infty}$ is symmetric. By Lemma 5.4.2, there is a non-zero solution $(\hat{Y_0}, \hat{V})$ of Eq.(5.4.24) corresponding to $\{y_n\}_{n=-\infty}^{\infty}$. Notice that if $(Y_0, 0)$ is a non-zero solution of Eq.(5.4.24), then $\{y_n\}_{n=-\infty}^{\infty}$ as defined by Eq.(5.4.23) and Eq.(5.4.21), is a periodic solution of Eq.(5.4.3) with period $k+1 = 2m$. Therefore, it follows from (b) of Lemma 5.4.3 that

$$\det(2EI - (B_0 + B_1 D + B_2)) \neq 0.$$

Consequently, the existence of $(\hat{Y_0}, \hat{V})$ implies that

$$\hat{V} \neq 0 \quad \text{and} \quad \det(\tilde{B_1}) = 0.$$

By the discussion about the linearization of Eq.(5.4.19), it follows from the Implicit Function Theorem that there exists $\alpha_0 > 0$ and a continuous function $X_0 = X_0(\alpha)$ from $(-\alpha_0, \alpha_0)$ into $\mathbf{R}^m$ such that $(X_0(\alpha), \alpha\hat{V})$ satisfies Eq.(5.4.19) for all $\alpha \in (-\alpha_0, \alpha_0)$ and $X_0(0) = (E, E, \ldots, E)^T \in \mathbf{R}^m$. Moreover one can write

$$X_0(\alpha) = (E, E, \ldots, E)^T + \alpha\hat{Y_0} + \alpha^2\hat{X_0}(\alpha), \tag{5.4.26}$$

where $\hat{X_0}(\alpha)$ is a continuous function from $(-\alpha_0, \alpha_0)$ to $\mathbf{R}^m$. By Lemma 5.4.1, $(X_0, W) = (X_0(\alpha), \alpha\hat{V})$ yields a symmetric periodic solution $\{x_n(\alpha)\}_{n=-\infty}^{\infty}$ of Eq. (5.4.1) with period $4m$ for each $\alpha \in (-\alpha_0, \alpha_0)$. Moreover it follows from Eq.(5.4.26) that one can write

$$x_n(\alpha) = E + \alpha y_n + \alpha^2\tilde{x}_n(\alpha) \quad \text{for} \quad n = 0, 1, \ldots \quad \text{and} \quad \alpha \in (-\alpha_0, \alpha_0)$$

where $\tilde{x}_n(\alpha)$, for $n = 0, 1, \ldots$, are continuous functions in $\alpha \in (-\alpha_0, \alpha_0)$. Consequently if $\{y_n\}_{n=-\infty}^{\infty}$ is of minimal period $4m$, then $\{x_n\}_{n=-\infty}^{\infty}$ is also of minimal period $4m$ for α near zero. The proof is complete. $\square$

Before we present some examples, we obtain the following consequence of Lemma 5.4.3.

Lemma 5.4.4 *Assume $k = 2m - 1$. Then the following statements are true.*

(a) Eq.(5.4.3) has a non-trivial (symmetric) periodic solution of period $4m$ if and only if the polynomials

$$\sum_{j=0}^{2m-2} b_j \lambda^{2m-1-j} \quad \text{and} \quad \lambda^{2m} + 1$$

have a common factor.

(b) If there is a solution $\lambda = e^{q\pi i/2m}$ of the equation

$$\sum_{j=0}^{2m-2} b_j \lambda^{2m-1-j} = 0$$

with q and $2m$ being relatively prime, then this λ is a solution of Eq.(5.4.25) and the corresponding symmetric periodic solution of Eq.(5.4.3) in (c) of Lemma 5.4.3 is of minimal period $4m$.

Proof. According to Lemma 5.4.3, Eq.(5.4.3) has a non-trivial (symmetric) periodic solution of period $4m$ if and only if there exists a solution $\lambda_0 = e^{2q\pi i/4m}$ (q integer) of Eq.(5.4.25). By (b) of Lemma 5.4.3, q is an odd integer. Therefore $\lambda_0^{2m} + 1 = 0$. This is equivalent to the statement that the polynomials $(\sum_{j=0}^{2m-2} b_j \lambda^{2m-1-j})$ and $(\lambda^{2m} + 1)$ have a common factor. From this equivalence, the last part of Lemma 5.4.4 is a consequence of part (c) of Lemma 5.4.3. $\qquad \square$

Next we will apply our result to some special cases of Eq.(5.4.1).
For $k = 2m - 1 = 3$, Eq.(5.4.1) becomes

$$x_{n+1} = \frac{a + b_0 x_n + b_1 x_{n-1} + b_0 x_{n-2}}{x_{n-3}}, n = 0, 1, \ldots \qquad (5.4.27)$$

where $b_2 = b_0$ by the symmetry. By Lemma 5.4.4 we compare the polynomial $(b_0 \lambda^3 + b_1 \lambda^2 + b_0 \lambda)$ and $(\lambda^4 + 1)$. Since

$$\lambda^4 + 1 = (\lambda^2 + \sqrt{2}\lambda + 1)(\lambda^2 - \sqrt{2}\lambda + 1),$$

these two polynomials have a common factor if and only if $b_1 = \sqrt{2}b_0$. $\lambda^2 + \sqrt{2}\lambda + 1 = 0$ has two solutions $\lambda = e^{3\pi i/4}$ and $\lambda = e^{5\pi i/4}$, which yield symmetric periodic solutions of Eq.(5.4.3) with minimal period 8 according to Lemma 5.4.4. Therefore, by Theorem 5.4.1 we have the following result

Theorem 5.4.2 *If $b_1 = \sqrt{2}b_0 > 0$ and $a > 0$, then there exist infinitely many symmetric periodic solutions of Eq.(5.4.27) with minimal period 8 near the positive equilibrium E of Eq.(5.4.27).*

For $k = 2m - 1 = 5$, Eq.(5.4.1) becomes

$$x_{n+1} = \frac{a + b_0 x_n + b_1 x_{n-1} + b_0 x_{n-2}}{x_{n-3}} \quad n = 0, 1, \ldots \quad (5.4.28)$$

where $b_3 = b_1$ and $b_4 = b_0$ for the symmetry. Since

$$\lambda^6 + 1 = (\lambda^2 + 1)(\lambda^2 + \sqrt{3}\lambda + 1)(\lambda^2 - \sqrt{3}\lambda + 1),$$

one can write

$$
\begin{aligned}
f(\lambda) &= b_0 \lambda^5 + b_1 \lambda^4 + b_2 \lambda^3 + b_1 \lambda^2 + b_0 \lambda \\
&= \lambda[(b_0 \lambda^2 + b_1 \lambda + b_2 - b_0)(\lambda^2 + 1) + (2b_0 - b_2)],
\end{aligned}
$$

or

$$f(\lambda) = \lambda[(b_0 \lambda^2 + (b_1 \mp \sqrt{3}b_0))(\lambda^2 \pm \sqrt{3}\lambda + 1) + (b_2 \mp \sqrt{3}b_1 + b_0)\lambda^2].$$

By Lemma 5.4.4, if $b_2 = 2b_0$ or $b_2 - \sqrt{3}b_1 + b_0 = 0$ then Eq.(5.4.3) has non-trivial symmetric periodic solutions of period 12. Observe that $\lambda^2 + \sqrt{3}\lambda + 1 = 0$ yields solutions $\lambda = e^{11\pi i/12}$ and $e^{13\pi i/12}$. Therefore it follows from Lemma 5.4.4 that if $b_2 - \sqrt{3}b_1 + b_0 = 0$, then there exist periodic solutions of Eq.(5.4.3) with minimal period 12. In view of the above, we have the following result:

Theorem 5.4.3 *Assume $b_1, b_2 \in [0, \infty)$ and $a > 0$. If $b_2 = 2b_0$ or $b_2 - \sqrt{3}b_1 + b_0 = 0$, then Eq.(5.4.28) has infinitely many symmetric periodic solutions of period 12 near the positive equilibrium E of Eq.(5.4.28). More precisely, if $b_2 - \sqrt{3}b_1 + b_0 = 0$, then Eq.(5.4.28) has infinitely many symmetric periodic solutions, each with minimal period 12 and arbitrarily near the positive equilibrium E.*

Research Project 5.4.1 *Extend the results of this section to equations where some of the coefficients are negative.*

Research Project 5.4.2 *Extend the results of this section to equations with complex coefficients.*

5.5   Notes

Our interest in the material of this chapter was influenced by Graham [1].

The results in Sections 5.1 - 5.3 are from Kocic, Ladas and Rodrigues [1]. Section 5.4 is from Cao and Ladas [1]. Some interesting results on periodic cycles are contained in Conway and Graham [1].

Chapter 6

Open Problems and Conjectures

Our aim in this chapter is to present some conjectures and some open problems about some interesting types of difference equations.

The conjectures are based on computer observations and deal with some simple looking equations with which we had a lot of fun, but lack of total success.

The open problems on the other hand are about equations which have not yet being thoroughly investigated and which, we hope, will stimulate a lot of interest and enthusiasm towards the development of some systematic results in this area.

6.1 The Rational Recursive Sequence $x_{n+1} = (a + bx_n)/(A + x_{n-1})$

Consider the rational recursive sequence

$$x_{n+1} = \frac{a + bx_n}{A + x_{n-1}}, \quad n = 0, 1, \dots \tag{6.1.1}$$

where

$$a, b, A \in (0, \infty) \tag{6.1.2}$$

and where the initial conditions x_{-1} and x_0 are arbitrary positive numbers. If $\bar{x}$ denotes the positive equilibrium of Eq.(6.1.1), then clearly

$$\bar{x} = \frac{a + b\bar{x}}{A + \bar{x}}$$

and so

$$\bar{x} = \frac{(b - A) + \sqrt{(b - A)^2 + 4a}}{2}. \tag{6.1.3}$$

On the basis of computer observations we pose the following conjecture:

153

Conjecture 6.1.1 *Assume that (6.1.2) holds. Then the poitive equilibrium $\bar{x}$ of Eq.(6.1.1) is globally asymptotically stable.*

In the special case where

$$b < A$$

and also in the special cases where

$$b \geq A \quad \text{with either} \quad a \leq Ab \quad \text{or} \quad Ab < a \leq 2A(b+A)$$

the conjecture has been confirmed. See Corollary 3.4.1(d) and (e). It is interesting to note that in view of Theorem 3.4.1(a), every solution of Eq.(6.1.1) with positive initial conditions is bounded away from zero and infinity by positive constants. In fact Eq.(6.1.1) is permanent.

The following lemma summarizes some properties of the semicycles of the positive solutions of Eq.(6.1.1) and may be useful in establishing the global asymptotic stability of $\bar{x}$.

Lemma 6.1.1 *Let $\{x_n\}$ be a nontrivial positive solution of Eq.(6.1.1). Then the following statements are true:*

(a) Every semicycle, except perhaps for the first one, has at least two terms. Furthermore, the second term in a positive semicycle, except for possibly the first, is greater than or equal to $\bar{x}$.

(b) The extreme in any semicycle occurs at either the first term or the second term. Furthermore after the first term, the remaining terms in a positive semicycle are strictly decreasing and the remaining terms in a negative semicycle are strictly increasing.

(c) In any two consecutive semicycles, their extrema cannot be consecutive terms.

(d) Assume that $Ab < a$. Then every semicycle, except for possibly the first, contains two or three terms. Furthermore, in every semicycle there is at most one term which follows the extreme.

Proof. We will give the proofs of (a), (b) and (d) for positive semicycles only. The proofs for negative semicycles are similar and will be omitted.
(a) Assume that for some n,

$$x_n < \bar{x} \quad \text{and} \quad x_{n+1} \geq \bar{x}.$$

Then

$$x_{n+2} = \frac{a + bx_{n+1}}{A + x_n} > \frac{a + b\bar{x}}{A + \bar{x}} = \bar{x}$$

from which the assertion follows.

(b) Assume that for some n the first two terms in a positive semicycle are x_n and x_{n+1}. Then $x_n \geq \bar{x}$, $x_{n+1} > \bar{x}$, and

$$\frac{x_{n+2}}{x_{n+1}} = \frac{1}{x_{n+1}} \frac{a + bx_{n+1}}{A + x_n} = \frac{\frac{a}{x_{n+1}} + b}{A + x_n} < \frac{\frac{a}{\bar{x}} + b}{A + \bar{x}} = 1$$

from which assertion follows.

(c) We will consider the case where a negative semicycle is followed by a positive one. The opposite case is similar and will be omitted. So assume for the sake of contradiction that for some n, the last two terms in a negative semicycle are x_{n-1} and x_n, with $x_{n-1} \geq x_n$, and the first two terms in the succeeding positive semicycle are x_{n+1} and x_{n+2} with $x_{n+1} \geq x_{n+2}$. Then

$$x_{n+2} \leq x_{n+1} = \frac{a + bx_n}{A + x_{n-1}} < \frac{a + bx_{n+1}}{A + x_n} = x_{n+2}$$

and this contradiction completes the proof.

(d) Set $g(x) = \frac{a+bx}{A+x}$ and observe that $g(x)$ is decreasing in x. Now assume that x_n, x_{n+1}, and x_{n+2} are three terms in a positive semicycle. Then

$$x_{n+3} = \frac{a + bx_{n+2}}{A + x_{n+1}} < \frac{a + bx_{n+1}}{A + x_{n+1}} = g(x_{n+1}) < g(\bar{x}) = \bar{x}$$

and the proof is complete. □

6.2 The Rational Recursive Sequence $x_{n+1} = (a + bx_n^2)/(c + x_{n-1}^2)$

Computer observations together with some analysis of the solutions will easily convince a curious reader that the solutions of the equation in the title possess an amazingly rich and surprising behavior. Before we state some open problems we present some interesting results for the special case

$$x_{n+1} = \frac{2x_n^2}{1 + x_{n-1}^2}, \quad n = 0, 1, \dots \tag{6.2.1}$$

where the initial conditions x_{-1} and x_0 are arbitrary positive numbers.

Clearly, every solution of Eq.(6.2.1) is positive. Also, Eq.(6.2.1) has the identically equal to 1 solution and, unless $x_{-1} = x_0 = 1$, no solution can be eventually equal to 1. It is easy to see that every solution of Eq.(6.2.1) is bounded from above. In fact the following is true.

Lemma 6.2.1 *Let $\{x_n\}$ be a solution of Eq.(6.2.1). Then*

$$x_n < 32 \quad \text{for} \quad n \geq 3. \tag{6.2.2}$$

Proof. Observe that for $n \geq 3$,

$$x_{n+3} = \frac{2x_{n+2}^2}{1 + x_{n+1}^2} = \cdots$$

$$= \frac{128x_n^8}{(1 + x_n^2)^2(1 + x_{n-1}^2)^2[(1 + x_{n-1}^2)^2 + 4x_n^4]}$$

$$< \frac{128x_n^8}{x_n^4 4x_n^4} = 32.$$

$\square$

The following theorem describes the asymptotic behavior of all solutions of Eq. (6.2.1).

Theorem 6.2.1 *Assume that $\{x_n\}$ is a solution of Eq.(6.2.1) which is neither identically equal to 1 nor strictly increasing to 1. Then*

$$\lim_{n \to \infty} x_n = 0. \tag{6.2.3}$$

Proof. Note that 1 is the only positive equilibrium of Eq.(6.2.1). Also $\{x_n\}$ is a positive and bounded sequence. Hence, there exists an $n_0 \geq 0$ such that

$$x_{n_0} \leq x_{n_0-1}.$$

(Otherwise $\{x_n\}$ would be strictly increasing to 1 or to ∞.) Now observe that

$$x_{n_0+1} = \frac{2x_{n_0}^2}{1 + x_{n_0-1}^2} \leq x_{n_0}$$

and by induction

$$x_{n+1} \leq x_n \quad \text{for} \quad n \geq n_0.$$

Hence

$$L = \lim_{n \to \infty} x_n$$

exists and either $L = 0$ or $L = 1$. Assume for the sake of contradiction that $L = 1$. Set

$$y_n = \frac{x_{n+1}}{x_n} \quad \text{for} \quad n \geq n_0.$$

Then for $n \geq n_0$, $0 < y_n \leq 1$ and

$$y_{n+1} = \frac{x_{n+2}}{x_{n+1}} = \frac{2x_{n+1}}{1 + x_n^2} \leq \frac{2x_{n+1}}{2x_n} = y_n.$$

Also

$$\lim_{n \to \infty} y_n = \frac{L}{L} = 1.$$

Hence $y_n = 1$ for $n \geq n_0$ which implies that $x_{n+1} = x_n$ for $n \geq n_0$. Therefore $x_n = 1$ for $n \geq n_0$ which is true only if $x_{-1} = x_0 = 1$. This is a contradiction and the proof is complete. □

Does Eq.(6.2.1) have a positive solution which is strictly increasing to 1? On the basis of some computer observations we had conjectured that Eq.(6.2.1) had no such solutions. Our conjecture was refuted by Northsfield [1] who established the following surprising result.

Theorem 6.2.2 *Eq.(6.2.1) has a solution which is strictly increasing to 1.*

Proof. Assume, for the sake of contradiction, that Eq.(6.2.1) has no positive solution which is strictly increasing to 1. Then we can easily see that for each solution $\{x_n\}$ of Eq.(6.2.1) with

$$0 < x_{-1} < x_0 < 1,$$

there exists a positive integer N (which depends on the solution) such that

$$x_{-1} < x_0 < \cdots \leq x_N \quad \text{and} \quad x_j > x_{j+1} \quad \text{for} \quad j \geq N.$$

Furthermore $x_N \neq 1$, for otherwise

$$1 > x_{N+1} = \frac{2}{1 + x_{N-1}^2} > 1$$

which is a contradiction.

Now for $1/2 \leq x \leq 1/\sqrt{2}$, define the sequence of continuous functions

$$f_{-1}(x) = x, \quad f_0(x) = 2x^2$$

and

$$f_{n+1}(x) = \frac{2f_n^2(x)}{1 + f_{n-1}^2(x)}, \quad n = 0, 1, \ldots.$$

For each $x \in [1/2, 1/\sqrt{2}]$, the sequence of positive numbers $\{f_n(x)\}$ is a solution of Eq.(6.2.1) and by Lemma 6.2.1 this sequence in bounded.

Let

$$s(x) = \sup_n f_n(x).$$

In view of the hypothesis that Eq.(6.2.1) has no solution which is strictly increasing to 1, it follows that for every $x \in [1/2, 1/\sqrt{2}]$, there exists a positive integer N (which depends on x) such that

$$s(x) = f_N(x), \quad f_{N-1}(x) \le f_N(x), \quad \text{and} \quad f_N(x) > f_{N+1}(x).$$

We will now prove that the function s is continuous for all $x \in [1/2, 1/\sqrt{2}]$. To this end let $x \in [1/2, 1/\sqrt{2}]$ and let

$$0 < \epsilon < \frac{f_N(x) - f_{N+1}(x)}{2}.$$

By the continuity of $f_0, \ldots, f_{N+1}$, choose $\delta > 0$ such that

$$|x' - x| < \delta \quad \text{implies} \quad \sup_{0 \le k \le N+1} |f_k(x') - f_k(x)| < \epsilon.$$

Note that

$$f_{N+1}(x') < f_{N+1}(x) + \epsilon < f_N(x) - \epsilon < f_N(x')$$

and so

$$s(x') = f_k(x') \quad \text{for some} \quad k \le N.$$

Therefore

$$\begin{aligned} s(x) - \epsilon &= f_N(x) - \epsilon < f_N(x') \le s(x') = \sup_{0 \le k \le N} f_k(x') \\ &= \sup_{0 \le k \le N} [f_k(x) + \epsilon] = s(x) + \epsilon \end{aligned}$$

and so

$$|s(x') - s(x)| < \epsilon$$

which establishes our claim that s i continuous on $[1/2, 1/\sqrt{2}]$.

As

$$s\left(\frac{1}{2}\right) = \frac{1}{2} \quad \text{and} \quad s\left(\frac{1}{\sqrt{2}}\right) > 1,$$

it follows that there exists an $x^* \in (1/2, 1/\sqrt{2})$ such that

$$s(x^*) = 1.$$

Hence there exists an N such that

$$f_{N-1}(x^*) \le 1, \quad f_N(x^*) = 1, \quad \text{and} \quad f_{N+1}(x^*) < 1.$$

But then

$$1 > f_{N+1}(x^*) = \frac{2}{1 + f_{N-1}^2(x^*)} \ge 1$$

which is a contradiction. The proof is complete   □

Open Problem 6.2.1 *Investigate the global asymptotic stability of the rational recursive sequence*

$$x_{n+1} = \frac{bx_n^2}{1 + x_{n-1}^2}, \ n = 0, 1, \ldots$$

where $b \in (0, \infty)$ and where the initial conditions x_{-1} and x_0 are arbitrary positive numbers.

Open Problem 6.2.2 *Investigate the global asymptotic stability and the periodic character of the solutions of the rational recursive sequence*

$$x_{n+1} = \frac{a + bx_n^2}{c + x_{n-1}^2}, \ n = 0, 1, \ldots$$

where $a, b \in [0, \infty)$ with $a + b > 0$ and where the initial conditions x_{-1} and x_0 are arbitrary positive numbers. Study the semicycles of the solution.

6.3   A Model for an Annual Plant

In this section we pose an open problem about the global asymptotic stability of the difference equation

$$x_{n+1} = \frac{\lambda x_n}{(1 + ax_{n-k})^p + b\lambda x_{n-m}}, \ n = 0, 1, \ldots \tag{6.3.1}$$

where

$$a, b, p \in (0, \infty), \ \lambda \in (1, \infty) \ \text{ and } \ k, m \in \{0, 1, \ldots\}. \tag{6.3.2}$$

Let $r = \max\{k, m\}$. We will assume that the initial conditions $x_{-r}, \ldots, x_0$ of Eq.(6.3.1) are arbitrary positive numbers.

Eq.(6.3.1), when $k = m = 0$, has been proposed by Watkinson [1], (see also Watkinson, Lonsdale and Andrew [1]) as a model describing the population dynamics of the annual plant *Sorgum intrans*. Also, when $p = 1$, Eq.(6.3.1) is a special case of the discrete delay logistic model with several delays. See Section 4.1.

Eq.(6.3.1) has a unique positive equilibrium $\bar{x}$. Furthermore $\bar{x}$ is the unique positive solution of the equation

$$\lambda = (1 + a x)^p + b\lambda x. \tag{6.3.3}$$

The following result gives necessary and sufficient conditions for the oscillation of all positive solutions of Eq.(6.3.1) about $\bar{x}$.

Theorem 6.3.1 *Assume that (6.3.2) holds and suppose that*

$$\frac{a\bar{x}p(a + a\bar{x})^{p-1}}{\lambda} + b\bar{x} + m + k \neq 1.$$

Then all positive solutions of Eq.(6.3.1) oscillate about $\bar{x}$ if and only if all solutions of the linear difference equation

$$z_{n+1} - z_n + \frac{a\bar{x}p(1 + a\bar{x})^{p-1}}{\lambda}z_{n-k} + b\bar{x}z_{n-m} = 0, \; n = 0, 1, \ldots \tag{6.3.4}$$

oscillate about zero.

Proof. Set $x_n = \bar{x}e^{y_n}$. Then Eq.(6.3.1) becomes

$$y_{n+1} - y_n + \ln\left(\frac{(1 + a\bar{x}e^{y_{n-k}})^p + b\lambda\bar{x}e^{y_{n-m}}}{\lambda}\right) = 0, \; n = 0, 1, \ldots. \tag{6.3.5}$$

Clearly, every positive solution of Eq.(6.3.1) ocillate about $\bar{x}$ if and only if every solution of Eq.(6.3.5) oscillate about zero. Consider the function

$$f(x, y) = \ln\left(\frac{(1 + a\bar{x}e^x)^p + b\lambda\bar{x}e^y}{\lambda}\right). \tag{6.3.6}$$

It is easy to see that $f(x, y) \geq 0$ for $x, y \in [0, \infty)$, $f(x, y) \leq 0$ for $x, y \in (-\infty, 0]$, and $f(x, x) = 0$ only if $x = 0$. Since

$$D_x f(0, 0) = \frac{(1 + a\bar{x})^p}{\lambda} \quad \text{and} \quad D_y f(0, 0) = b\bar{x},$$

it follows that Eq.(6.3.4) is the linearized equation associated with Eq.(6.3.5). Therefore, by Theorem 1.2.4, it remains to show that there exists $\delta > 0$ such that

$$f(x, y) \geq \frac{(1 + a\bar{x})^p}{\lambda}x + b\bar{x}y \quad \text{for} \quad x, y \in [-\delta, 0].$$

To this end observe that the function

$$g(x, y) = f(x, y) - \frac{(1 + a\bar{x})^p}{\lambda}x + b\bar{x}y$$

has a local minimum equal to 0 at the point $(0, 0)$. Hence there exists $\delta > 0$ such that $g(x, y) \geq 0$ for $(x, y) \in [-\delta, 0]$. The proof is complete. $\square$

The following result gives some explicit sufficient conditions for the local asymptotic stability of the equilibrium $\bar{x}$ of Eq.(6.3.1).

Lemma 6.3.1 *Assume that (6.3.2) holds. Then the positive equilibrium $\bar{x}$ of Eq.* *(6.3.1) is locally asymptotically stable if one of the following conditions is satisfied:*

(a) $k = m$ and

$$\lambda > \frac{(1 + a\bar{x})^{p-1}(a\bar{x}(p-1) - 1)}{2 \cos \dfrac{k\pi}{2k+1}};$$

(b) $k = 0$, $m = 1$ and

$$\lambda > \frac{(1 + a\bar{x})^{p-1}(a\bar{x}(p-1) - 1)}{3} \quad \text{and} \quad b\bar{x} < 1;$$

(c) $k = 1$, $m = 0$ and

$$\lambda > \max\{(1 + a\bar{x})^{p-1}(1 + a\bar{x}(1 - p)), \ a\bar{x}p(1 + a\bar{x})^{p-1}\};$$

(d) $k = 0$, $m > 1$ and

$$(1 + a\bar{x})^{p-1}(a\bar{x}(p-1) - 1) < \lambda < \frac{a}{b}(1 + a\bar{x})^{p-1};$$

(e) $k > 1$, $m = 0$ and

$$\lambda > \max\{(1 + a\bar{x})^{p-1}(a\bar{x}(p-1) - 1), \ \frac{a}{b}(1 + a\bar{x})^{p-1}\};$$

(f) $k, m > 0$, $k \neq m$ and

$$(m - 1)\lambda < (1 + a\bar{x})^{p-1}(a\bar{x}(m - pk) + 1).$$

Proof. Clearly Eq.(6.3.4) is the linearized equation associated with Eq.(6.3.1). Condition (a) follows from Theorem 1.3.6. Conditions (b) and (c) are obtained by applying Theorem 1.3.4 and conditions (d) and (e) follow from Theorem 1.3.7. Finally condition (f) is obtained by applying Remark 1.2.2. □

The global asymptotic stability of the equilibrium $\bar{x}$ of Eq.(6.3.1) has not been investigated yet. We therefore pose the following open problem.

Open Problem 6.3.1 *Obtain conditions for the global asymptotic stability of the equilibrium $\bar{x}$ of Eq.(6.3.1).*

6.4   The Dynamics of $x_{n+1} = Ax_n^{-2} + x_{n-1}^{-1/2}$

Consider the difference equation

$$x_{n+1} = \frac{A}{x_n^2} + \frac{1}{\sqrt{x_{n-1}}}, \quad n = 0, 1, \ldots \tag{6.4.1}$$

where $A \in (0, \infty)$ and $x_{-1}, x_0 \in (0, \infty)$. Let $\bar{x}$ denote the unique positive equilibrium of Eq.(6.4.1). Our aim in this section is to pose the following conjecture which is predicted by computer observations.

Conjecture 6.4.1 *(a) Show that when*

$$0 < A < \frac{15}{4} \tag{6.4.2}$$

the positive equilibrium of Eq.(6.4.1) is globally asymptotically stable.

(b) Show that when

$$A > \frac{15}{4} \tag{6.4.3}$$

there exists a periodic cycle with period two which is globally asymptotically stable.

Although we are unable to establish the above conjecture, we have proven the following result.

Theorem 6.4.1 *(a) Assume that (6.4.2) holds. Then the positive equilibrium $\bar{x}$ of Eq.(6.4.1) is locally asymptotically stable.*

(b) Assume that (6.4.3) holds. Then Eq.(6.4.1) has a periodic solution with period two.

Proof. (a) Set $\bar{x} = \rho^2$. Then the linearized equation of Eq.(6.4.1) is

$$y_{n+1} + \frac{2A}{\rho^6} y_n + \frac{1}{2\rho^3} y_{n-1} = 0, \quad n = 0, 1, \ldots .$$

By Theorem 1.3.4 this equation is asymptotically stable if and only if

$$\frac{2A}{\rho^6} < 1 + \frac{1}{2\rho^3} < 2. \tag{6.4.4}$$

Note that the equilibrium $\bar{x} = \rho^2$ satisfies the equation

$$\rho^2 = \frac{A}{\rho^4} + \frac{1}{\rho} \tag{6.4.5}$$

where ρ is the unique positive root of the equation

$$t^6 - t^3 - A = 0.$$

Clearly $\rho > 1$, and so in view of (6.4.5), (6.4.4) is satisfied if and only if

$$\rho > \left(\frac{2A}{3}\right)^{1/3}. \tag{6.4.6}$$

Set

$$f(t) = t^6 - t^3 - A$$

and observe that

$$f(t) < 0 \quad \text{if} \quad 0 < t < \rho$$

and

$$f(t) > 0 \quad \text{if} \quad t > \rho.$$

Hence (6.4.6) is equivalent to $f((2A/3)^{1/3}) < 0$; that is,

$$A < \frac{15}{4}.$$

(b) Eq.(6.4.1) has a periodic solution of the form

$$\{p, q, p, q, \ldots\} \quad \text{or} \quad \{q, p, q, p, \ldots\}$$

if and only if

$$p = \frac{A}{q^2} + \frac{1}{\sqrt{p}} \quad \text{and} \quad q = \frac{A}{p^2} + \frac{1}{\sqrt{q}}. \tag{6.4.7}$$

Set $x = \sqrt{p}$ and $y = \sqrt{q}$. Then the system of algebraic equations (6.4.7) is equivalent to

$$\left.\begin{array}{rcl} x^2 & = & \dfrac{A}{y^4} + \dfrac{1}{x} \\[2mm] y^2 & = & \dfrac{A}{x^4} + \dfrac{1}{y} \end{array}\right\} \quad \text{with} \quad x, y > 0. \tag{6.4.8}$$

Set

$$\xi = x + y, \quad \eta = xy, \quad \text{and} \quad \zeta = \eta^3.$$

Then x and y are the roots of the quadratic equation

$$\lambda^2 - \xi\lambda + \eta = 0$$

and these roots are real, positive and distinct if and only if

$$\xi, \eta \in (0, \infty) \quad \text{and} \quad \eta < \frac{1}{4}\xi^2. \tag{6.4.9}$$

Cancel the denominators in (6.4.8) , then multiply the first equation by x and the second by y, equate the terms $x^4 y^4$, and divide by $x - y$. This leads to

$$A\xi = \eta(\xi^2 - \eta). \tag{6.4.10}$$

Cancel the denominators in (6.4.8), subtract and then divide by $x - y$. This yields

$$\eta^3 = -A + \xi(\xi^2 - 2\eta). \tag{6.4.11}$$

Subtract from the first equation in (6.4.8) the second, and use (6.4.11) to obtain

$$\xi = \frac{(A-1)\eta^3 + A^2}{\eta^4}. \tag{6.4.12}$$

By substituting (6.4.12) into (6.4.10) we find

$$G(\zeta) = \zeta^3 + (A-1)\zeta^2 + A^2(2-A)\zeta - A^4 = 0. \tag{6.4.13}$$

Note that

$$G(\zeta) < 0 \text{ if } z < \zeta \quad \text{and} \quad G(\zeta) > 0 \text{ if } z > \zeta. \tag{6.4.14}$$

In view of (6.4.9) and (6.4.12) we obtain

$$4\zeta^3 < (A-1)\zeta^2 + 2A^2(A-1)\zeta + A^4.$$

and so by using (6.4.13) we find

$$H(\zeta) = (A+3)(A-1)\zeta^2 + 2A^2(3-A)\zeta - 3A^4 > 0.$$

The positive root of this quadratic equation is

$$\zeta_1 = \frac{3A^2}{A+3}$$

and so $H(\zeta) > 0$ if and only if $\zeta > 3A^2/(A+3)$ which in view of (6.4.14) is true if and only if $G(3A^2/(A+3)) < 0$ that is

$$A > \frac{15}{4}.$$

The proof of the theorem is complete.   □

A related difference equation is

$$x_{n+1} = \frac{a}{x_n^2} + \frac{1}{x_{n-1}} \quad n = 0, 1, \ldots \tag{6.4.15}$$

where $A \in (0, \infty)$ and the initial conditions $x_{-1}, x_0 \in (0, \infty)$.

One can show that the following result is true.

Theorem 6.4.2 *The following statements are true:*

a) The unique positive equilibrium $\bar{x}$ of Eq.(6.4.15) is locally asymptotically stable if

$$a < 2\sqrt{3} \tag{6.4.16}$$

and unstable if

$$a > 2\sqrt{3}. \tag{6.4.17}$$

b) When (6.4.17) holds, Eq.(6.4.15) has a periodic solution with period two, $\{p, q, p, q, \ldots\}$. Furthermore

$$p = \frac{a + \sqrt{a^2 + 2 - 2\sqrt{1 + 4a^2}}}{2} \quad \text{and} \quad q = \frac{a - \sqrt{a^2 + 2 - 2\sqrt{1 + 4a^2}}}{2}.$$

Open Problem 6.4.1 *(a) For what values of a is the positive equilibrium $\bar{x}$ of Eq. (6.4.15) globally asymptotically stable?*

(b) For what values of a is the periodic solution $\{p, q, p, q, \ldots\}$ of Eq.(6.4.15) asymptotically stable? What is its basin of attraction?

Open Problem 6.4.2 *Consider the difference equation*

$$x_{n+1} = \frac{A}{x_n^2} + \frac{1}{x_{n-k}^p}, \quad n = 0, 1, \ldots \tag{6.4.18}$$

where

$$A, p \in (0, \infty) \quad \text{and} \quad k \in \{0, 1, \ldots\}$$

and the initial conditions $x_{-k}, \ldots, x_0$ are arbitrary positive numbers.

a) Obtain conditions on A, p and k under which the positive equilibrium of Eq. (6.4.18) is globally asymptotically stable?

b) Obtain conditions on A, p and k under which Eq.(6.4.18) has periodic cycles of period two. Under what conditions on A, p and k are these periodic cycles stable? What is their basin of attraction?

c) Do there exist periodic cycles of period greater than two?

6.5   A Discrete Model with Quadratic Nonlinearity

In this section we pose several research projects concerning the behavior of the positive solutions of the difference equation

$$N_{n+1} - N_n = N_n(a + bN_{n-k} - cN_{n-k}^2), \quad n = 0, 1, \ldots \qquad (6.5.1)$$

where

$$a, c \in (0, \infty), \quad b \in (-\infty, \infty), \quad \text{and} \quad k \in \{0, 1, \ldots\}. \qquad (6.5.2)$$

Eq.(6.5.1) may be viewed as a discrete analogue of the delay differential equation

$$\dot{N}(t) = N(t)[a + bN(t - \tau) - cN(t - \tau)^2], \quad t \geq 0 \qquad (6.5.3)$$

where

$$a, c \in (0, \infty), \quad b \in (-\infty, \infty), \quad \text{and} \quad \tau \in [0, \infty). \qquad (6.5.4)$$

Eq.(6.5.3), which is a model of single species with a quadratic per-capita growth rate, exhibits the characteristic feature that at high densities, intraspecific competition dominates while when $b > 0$, at low densities, intraspecific mutualism dominates. The same feature is, clearly, exhibited by Eq.(6.5.1).

The differential equation (6.5.3) was investigated by Gopalsamy and Ladas [1].

The oscillation of Eq.(6.5.1) and the global attractivity of the positive equilibrium when $k = 0$ were investigated by Rodrigues [1].

Let $\bar{N}$ denote the unique positive equilibrium of Eq.(6.5.1). Then

$$\bar{N} = \frac{b + \sqrt{b^2 + 4ac}}{2c}.$$

By applying linearized stability analysis (that is by using Theorems 1.3.5 and 1.3.6) we obtain the following result.

Theorem 6.5.1 *Assume that (6.5.2) holds. Then $\bar{N}$ is locally asymptotically stable provided that*

$$\bar{N}\sqrt{b^2 + 4ac} < 2\cos\frac{k\pi}{2k + 1}.$$

By applying the oscillation invariant transformation

$$N_n = \bar{N}e^{x_n},$$

Eq.(6.5.1) reduces to

$$x_{n+1} - x_n + pf(x_{n-k}) = 0 \qquad (6.5.5)$$

where

$$p = \bar{N}(2c\bar{N} - b)$$

and

$$f(u) = -\frac{\ln(a + 1 + b\bar{N}e^u - c\bar{N}^2 e^{2u})}{\bar{N}(2c\bar{N} - b)}.$$

The following result is now a consequence of Theorem 1.2.3 and Corollary 1.2.1.

Theorem 6.5.2 *Every eventually positive solution of Eq.(6.5.1) oscillates about $\bar{N}$ if and only if one of the following conditions holds:*

a) $k = 0$ and $\bar{N}(2c\bar{N} - b) > 1$;

b) $k \geq 1$ and $\bar{N}(2c\bar{N} - b) > \dfrac{k^k}{(k+1)^{k+1}}$.

Research Project 6.5.1 *Obtain a global stability result for the positive equilibrium $\bar{N}$ of Eq.(6.5.1) when $k \geq 1$.*

Research Project 6.5.2 *Obtain necessary and sufficient conditions for the oscillation of all positive solutions of the difference equation*

$$N_{n+1} - N_n = N_n(a + bN_{n-k} - cN_{n-l}^2), \quad n = 0, 1, \ldots \qquad (6.5.6)$$

where

$$a, c \in (0, \infty), \quad b \in (-\infty, \infty), \quad and \quad k, l \in \{0, 1, \ldots\}. \qquad (6.5.7)$$

Research Project 6.5.3 *Assume that (6.5.7) holds. Obtain a global asymptotic stability result for the positive solutions of Eq.(6.5.6).*

Research Project 6.5.4 *Assume that (6.5.2) holds. Obtain explicit sufficient conditions on a, b, c and k so that all solutions of Eq.(6.5.1) with appropriate initial conditions remain positive for all $n \geq 0$.*

6.6   A Logistic Equation with Piecewise Constant Argument

The study of equations with piecewise constant arguments was originated by Wiener and his collaborators. See, for example, Aftabizadeh and Wiener [1, 2] Aftabizadeh, Wiener and Xu [1] and Cooke and Wiener [1-3]. These equations provide a rich source for obtaining difference equations of interesting types. One of these equations was investigated in Section 4.8. For some recent results on logistic equations with piecewise constant arguments see Carvalho and Cooke [1], Gopalsamy, Kulenović and Ladas [1], Huang, Y. K. [1], and Norris and Soewono [1].

In this section we will analyze the logistic equation with piecewise constant argument

$$\dot{x}(t) = \mu x(t)(1 - x([\frac{t+1}{2}])), \ t \geq 0 \tag{6.6.1}$$

where $\mu \in (0, \infty)$.

By a solution of Eq.(6.6.1) we mean a function $x(t)$ which satisfies the following properties:

(a) $x \in C[[0, \infty), \mathbf{R}]$;

(b) The derivative $\dot{x}(t)$ exists at each point $t \in [0, \infty)$, with the possible exception at the points $t = 2n + 1$ for $n = 0, 1, \ldots$ where finite one sided derivatives exist;

(c) Eq.(6.6.1) is satisfied in each interval $[2n - 1, 2n + 1) \cap [0, \infty)$ for $n = 0, 1, \ldots$.

Let y_0 be a given real number. Then by working in the interval $2n - 1 \leq t \leq 2n + 1$ for $n = 0, 1, \ldots$, we can see that Eq.(6.6.1) has a unique solution $x(t)$ satisfying the initial condition

$$x(0) = x_0. \tag{6.6.2}$$

Furthermore

$$x(t) = x_{2n} e^{\mu(1-x(n))(t-2n)} \ \text{ for } \ t \in [2n - 1, 2n + 1) \cap [0, \infty).$$

Set $e^{y_n} = x(n)$. Then by using the continuity of the solution at the end points of the interval $[2n - 1, 2n + 1)$, it follows that $\{y_n\}$ satisfies the system of difference equations

$$\left. \begin{array}{rlr} y_{2n+1} & = & y_{2n} + \mu(1 - e^{y_n}), \quad n = 0, 1, \ldots \\ \\ y_{2n} & = & y_{2n-1} + \mu(1 - e^{y_n}), \quad n = 1, 2, \ldots \end{array} \right\} \tag{6.6.3}$$

The linearized version of this system,

$$\left. \begin{array}{rlr} a_{2n+1} & = & a_{2n} - \mu a_n, \quad n = 0, 1, \ldots \\ \\ a_{2n} & = & a_{2n-1} - \mu a_n, \quad n = 1, 2, \ldots \end{array} \right\} \tag{6.6.4}$$

was investigated by Georgiou, Ladas and Vlahos [1]. They showed that every solution of (6.6.4) oscillates about zero if and only if $\mu > 0$. They also showed that when $\mu > 1$ every nontrivial solution of (6.6.4) is unbounded.

One can easily see that for the nonlinear system (6.6.1) every solution oscillates about zero if and only if $\mu > 0$.

Conjecture 6.6.1 *Assume that $y_0 \in (0, \infty)$. Show that every solution of (6.6.3) is unbounded.*

6.7 Discrete Epidemic Models

Our aim in this section is to pose some research projects about the behavior of solutions of the deterministic epidemic models

$$x_{n+1} = (1 - x_n)(1 - e^{-Ax_n}), \quad n = 0, 1, \ldots \tag{6.7.1}$$

$$x_{n+1} = (1 - x_n - x_{n-1})(1 - e^{-Ax_n}), \quad n = 0, 1, \ldots \tag{6.7.2}$$

$$x_{n+1} = (1 - x_n - x_{n-2} - x_{n-3})(1 - e^{-Ax_n}), \quad n = 0, 1, \ldots \tag{6.7.3}$$

and generally,

$$x_{n+1} = (1 - \sum_{j=0}^{k-1} x_n)(1 - e^{-Ax_n}), \quad n = 0, 1, \ldots. \tag{6.7.4}$$

The above equations are special cases of epidemic models which were derived in Beddington, Free and Lawton [1] and Cooke, Calef and Level [1].

Research Project 6.7.1 *Investigate the oscillatory behavior, the global asymptotic stability and the periodic character of solutions of Eqs. (6.7.1) - (6.7.4).*

Research Project 6.7.2 *How should we choose the initial conditions of Eq.(6.7.4) so that the solutions remain positive for all $n = 0, 1, \ldots$?*

6.8 Volterra Difference Equations

Consider the linear Volterra difference equation

$$x_{n+1} = ax_n + \sum_{j=0}^{n} b_{n-j}x_j, \quad n = 0, 1, \ldots \tag{6.8.1}$$

where

$$a, b_0, b_1, \ldots \in \mathbf{R} \quad \text{and} \quad \sum_{i=0}^{\infty} |b_i| < \infty. \tag{6.8.2}$$

If

$$Z(a_n) = \sum_{n=0}^{\infty} \frac{a_n}{z^n}$$

denotes the z-transform of the sequence $\{a_n\}$, then by taking z-transforms on both sides of Eq.(6.8.1) we obtain

$$Z(x_n) = \frac{z x_0}{z - a - Z(b_n)}.$$

It is known, see Elaydi [1], that Eq.(6.8.1) is asymptotically stable if and only if the complex function

$$g(z) = z - a - Z(b_n)$$

has no zeros in $|z| \geq 1$. By utilizing this result, Elaydi [1] obtained conditions for the asymptotic stability of Eq.(6.8.1) which are explicitly given in terms of a and $\{b_n\}$. See also Babai and Lengyel [1], Crisci, Jackiewicz, Russo, and Vecchio [1], Elaydi [2], Elaydi and Kocic [1], Elaydi and Zhang [1-2], Kiventidis [1], Ladas, Philos and Sficas [1], and Lakshmikantham, Matrosov and Sivasundaram [1].

Open Problem 6.8.1 Nonlinear Volterra difference equations
Obtain stability and oscillation results for the nonlinear Volterra difference equations

$$x_{n+1} - x_n + \sum_{j=0}^{n} b_{n-j}(e^{x_j} - 1) = 0, \ n = 0, 1, \ldots \tag{6.8.3}$$

and

$$y_{n+1} = y_n \left(a - \sum_{j=-\infty}^{n} b_{n-j} y_j \right), \ n = 0, 1, \ldots. \tag{6.8.4}$$

6.9 Global Attractivity of
$x_{n+1} - x_n + p x_{n-k} = f(x_{n-m})$

Consider the nonlinear difference equation

$$x_{n+1} - x_n + p x_{n-k} = f(x_{n-m}), \ n = 0, 1, \ldots \tag{6.9.1}$$

where k and m are nonnegative integers, $p \in (0, \infty)$,

$$f \in C[\mathbf{R}, \mathbf{R}], \quad \text{and} \quad f(u)u \geq 0 \quad \text{for} \quad u \in \mathbf{R}. \tag{6.9.2}$$

The following result of Györi, Ladas and Vlahos [1] (see also Györi and Ladas [2]), gives sufficient conditions for the global attractivity of the zero solution of Eq.(6.9.1).

Theorem 6.9.1 *Assume that (6.9.2) holds and that*

$$0 < p < \frac{k^k}{(k+1)^{k+1}}. \tag{6.9.3}$$

Then the following statements are equivalent:

(a) Every solution of Eq.(6.9.1) tends to zero as $n \to \infty$.

(b) $|f(u)| < p|u|$ for all $u \neq 0$.

A detailed proof of Theorem 6.9.1 can be found in Györi and Ladas [1,2].

The question of the global attractivity of the zero equilibrium of Eq.(6.9.1) when condition (6.9.3) is not satisfied is still open. We therefore pose the following problem.

Open Problem 6.9.1 *Assume that (6.9.2) holds and that*

$$p \geq \frac{k^k}{(k+1)^{k+1}}. \tag{6.9.4}$$

Find necessary and sufficient conditions for the global attractivity of the zero solution of Eq.(6.9.1).

A corollary of Theorem 6.9.1 is the following global stability result for the linear equation

$$x_{n+1} - x_n + px_{n-k} - qx_{n-m} = 0, \; n = 0, 1, \ldots. \tag{6.9.5}$$

Corollary 6.9.1 *Assume that k and m are nonnegative integers, $q \in [0, \infty)$, and that condition (6.9.3) holds. Then the zero solution of the linear difference equation (6.9.5) is globally asymptotically stable if and only if*

$$p > q. \tag{6.9.6}$$

The following result gives sufficient conditions for the global stability of the zero solution of Eq.(6.9.5) when (6.9.3) is not satisfied.

Theorem 6.9.2 *Assume that k and m are nonnegative integers, $p \in (0, \infty)$, and $q \in [0, \infty)$. Suppose also that*

$$kp < 1 \quad \text{and} \quad q < p\frac{1 - kp}{1 + kp}. \tag{6.9.7}$$

Then the zero solution of Eq.(6.9.5) is globally asymptotically stable.

Proof. By introducing the substitution

$$x_n = u_n + p\sum_{i=1}^{k} u_{n-i} \quad \text{for} \quad n = -k, -k+1, \dots$$

with $u_{-2k} = \cdots = u_{-k-1} = 0$, we see that for $n \geq 0$,

$$u_{n+1} - (1-p)u_n + p^2 \sum_{i=1}^{k} u_{n-k-i} - qu_{n-m} - pq\sum_{i=1}^{k} u_{n-m-i} = 0. \tag{6.9.8}$$

The result is now a consequence of Remark 1.2.1 and the observation that (6.9.7) implies that

$$|1 - p| + kp^2 + q + kpq < 1.$$

The proof is complete.   □

For the linear equation (6.9.5) we formulate the following problem:

Open Problem 6.9.2 *Assume that (6.9.4) holds. Find explicit necessary and sufficient conditions for the global asymptotic stability of the zero solution of Eq.(6.9.5).*

6.10   Neural Networks

The difference equations

$$X(t+1) \;=\; AX(t)[1 - X(t)], \tag{6.10.1}$$

$$X(t+1) \;=\; AX(t)[1 - X(t) - X(t-1)], \tag{6.10.2}$$

$$X(t+1) \;=\; AX(t)[1 - X(t) - X(t-1) - X(t-2)], \tag{6.10.3}$$

are cited by Morimoto [1,2] as appropriate recurrence equations for the activity $X(t)$ of mutually connected neural networks for refractory periods of 1, 2, and 3, respectively. Eq.(6.10.1) is the well-known logistic equation and has been extensively investigated in the literature. See, for example, Collet and Eckman [1], Devaney [1], Hale and Kocak [1], and May [2, 3, 6].

The bifurcation diagram of Eqs.(6.10.2) and (6.10.3) were studied by Morimoto [1, 2].

A search for the existence of periodic solutions for Eqs.(6.10.2) and (6.10.3) was made by Brown [1,2]. He showed that for $A > 0$, Eq.(6.10.2) has no solutions with prime period 2 or 3 although solutions with period 4 can occur. For Eq.(6.10.3) Brown showed that for $A > 0$, apart from equilibrium solutions, there are no stable periodic solutions. However, for $A < 0$, there are stable solutions with period 2 and unstable solutions of period 5 but no solutions with period 3, 4 and 6.

Persistence of Eq.(6.10.2) was invetigated by Levine, Scudo and Plunkett [1]. In Kocic and Ladas [2] the oscillation of the more general equation

$$x_{n+1} = Ax_n(1 - \sum_{i=0}^{m} B_i x_{n-i}), \quad n = 0, 1, \ldots \qquad (6.10.4)$$

was investigated. By employing the linearized oscillation result which is described by Theorem 1.2.4, Kocic and Ladas [2] proved the following result.

Theorem 6.10.1 *Assume that* $m \in \{0, 1, \ldots\}$, $A \in (1, \infty)$, *and* $B_0, \ldots, B_m \in (0, \infty)$ *and that*

$$A < \frac{4B_0}{\sum\limits_{i=0}^{m} B_i}. \qquad (6.10.5)$$

Let $\bar{x}$ *be the unique positive equilibrium of Eq.(6.10.4). Then every positive solution of Eq.(6.10.4) with initial conditions*

$$x_{-m}, \ldots, x_0 \in (0, \frac{A\bar{x}}{A-1})$$

oscillates about $\bar{x}$ *if and only if the equation*

$$\lambda - 1 + A\bar{x} \sum_{i=0}^{m} B_i \lambda^{-i} = 0$$

has no positive roots.

The global stability of Eqs.(6.10.2) and (6.10.3) and the global stability and periodicity of the Eq.(6.10.4) with $m \geq 2$ remain open questions at this time.

Open Problem 6.10.1 *Assume that $A \in (1, \infty)$. Obtain conditions for the global asymptotic stability of the positive equilibrium of Eqs.(6.10.2) and (6.10.3).*

Open Problem 6.10.2 *Assume that (6.10.5) holds with $m \geq 3$. Obtain conditions for the global asymptotic stability of the positive equilibrium of Eq.(6.10.4).*

Open Problem 6.10.3 *Assume that $A \in (1, \infty)$. Obtain explicit conditions on the parameters $A, B_0, \ldots, B_m$, and m so that all solutions of Eq.(6.10.4) with appropriate initial conditions remain positive for all $n \geq 0$.*

6.11   The Fibonacci Sequence Modulo π

Our aim in this note is to show how we arrived in a totally unexpected way at the Fibonacci sequence modulo π

$$F_{n+2} = (F_{n+1} + F_n) \bmod \pi \quad \text{for} \quad n = 0, 1, \ldots \tag{6.11.1}$$

and to pose the study of this, as yet unexplored equation, as a research problem.

Newton's method for solving the equation $f(x) = 0$ is

$$x_{n+1} = x_n - \frac{f(x_n)}{f'(x_n)}, \ n = 0, 1, \ldots.$$

If we replace $f'(x_n)$ by $(f(x_n) - f(x_{n-1}))/(x_n - x_{n-1})$, we obtain the delay difference equation

$$x_{n+1} = \frac{x_{n-1} f(x_n) - x_n f(x_{n-1})}{f(x_n) - f(x_{n-1})}, \ n = 0, 1, \ldots.$$

For the function

$$f(x) = x^2 - p$$

this reduces to

$$x_{n+1} = \frac{x_n x_{n-1} + p}{x_n + x_{n-1}}, \ n = 0, 1, \ldots. \tag{6.11.2}$$

One can see that for $p \geq 0$ and for arbitrary positive initial conditions x_0 and x_1,

$$\lim_{n \to \infty} x_n = \sqrt{p}.$$

For $p = -1$ Eq.(6.11.2) becomes,

$$x_{n+1} = \frac{x_n x_{n-1} - 1}{x_n + x_{n-1}}, \ n = 0, 1, \ldots.$$

Open Problem 6.11.1 The Secant Method for Finding $\sqrt{-1}$

Study the asymptotic behavior and the periodic nature of solutions of the difference equation

$$y_{n+2} = \frac{y_n y_{n+1} - 1}{y_n + y_{n+1}}, \quad n = 0, 1, \ldots \tag{6.11.3}$$

where y_0 and y_1 are arbitrary real numbers such that y_n exists for all n.

Let $\{y_n\}_{n=0}^\infty$ be a solution of Eq.(6.11.3) and set

$$y_n = \cot z_n \quad \text{for} \quad n = 0, 1, \ldots$$

where

$$z_n \in (0, \pi) \quad \text{for} \quad n = 0, 1, \ldots.$$

Then

$$\cot z_{n+2} = \frac{\cot z_n \cot z_{n+1} - 1}{\cot z_n + \cot z_{n+1}} = \cot(z_{n+1} + z_n)$$

and so

$$z_{n+2} = (z_{n+1} + z_n) \bmod \pi \quad \text{for} \quad n = 0, 1, \ldots \tag{6.11.4}$$

where

$$z_0 = \cot^{-1} y_0 \quad \text{and} \quad z_1 = \cot^{-1} y_1.$$

Eq.(6.11.4) is the **Fibonacci sequence modulo π**.

The interested reader may consult the paper Wall [1] on Fibonacci numbers modulo an integer m,

$$x_{n+1} = (x_n + x_{n-1}) \bmod m. \tag{6.11.5}$$

See also Adler [1], Andreassian [1], Burke and Webb [1], Chang [1], Ehrhart [1], Enrlich [1], Gill and Miller [1], Knuth [1], Mamangakis [1], Morgan [1], Nagasaka [1], Robinson [1], Vince [1], Vinson [1], Waddill [1], and Yalavigi [1].

When m is not an integer, the study of Eq.(6.11.5) remains an interesting open problem. In particular we pose the following problem when $m = \pi$.

Open Problem 6.11.2 The Fibonacci Sequence Modulo π

Assume that z_0 and z_1 are arbitrary numbers in $(0, \pi)$. Study the asymptotic behavior and the periodic character of solutions of the difference equation (6.11.4).

For a connection between Eq.(6.11.4) and Fibonacci numbers when $p = 0$, see Thoro [1]. For a connection between the secant method for the equation $x^2 - x - 1 = 0$ and ratios of Fibonacci numbers, see Gill and Miller [1].

6.12   Notes

Lemma 6.1.1 is from Jaroma [1]. The results in Section 6.2 are from Camouzis, Ladas and Rodrigues [1] and Northsfield [1]. The results in Section 6.3 are new. The results in Section 6.4 are from Arciero, Ladas and Schultz [1]. Theorems 6.5.1 and 6.5.2 are from Rodrigues [1]. Theorem 6.9.2 is new. The conjectures in Section 6.11 are from Grove, Kocic and Ladas [1].

Appendix A

The Riccati Difference Equation

In this appendix we study the properties of solutions of the so called **Riccati differ-ence equation**,

$$x_{n+1} = \frac{a_n x_n + b_n}{c_n x_n + d_n}, \ n = 0, 1, \ldots \qquad (A.1)$$

where $\{a_n\}$, $\{b_n\}$, $\{c_n\}$, $\{d_n\}$, are given sequences of real numbers, such that

$$a_n d_n - b_n c_n \neq 0 \ \text{ and } \ c_n \neq 0 \ \text{ for } \ n = 0, 1, \ldots. \qquad (A.2)$$

In the sequel we assume without further mention that the initial condition x_0 is chosen such that the solution $\{x_n\}$ exists for all positive integers n.

The following theorem summarizes several properties of the solutions of the Riccati difference equation Eq.(A.1).

Theorem A.1 *Assume that (A.2) holds. Then the following statements are true:*

(a) The transformation

$$c_n x_n + d_n = \frac{y_{n+1}}{y_n} \ \text{ for } \ n = 0, 1, \ldots \ \text{ with } \ y_0 = 1 \qquad (A.3)$$

reduces Eq.(A.1) to the linear second-order equation

$$y_{n+2} = p_n y_{n+1} + q_n y_n \ n = 0, 1, \ldots \qquad (A.4)$$

with

$$y_0 = 1, \quad y_1 = c_0 x_0 + d_0$$

where

$$p_n = \frac{c_n d_{n+1} + a_n c_{n+1}}{c_n} \ \text{ and } \ q_n = (b_n c_n - a_n d_n)\frac{c_{n+1}}{c_n}. \qquad (A.5)$$

177

(b) Let $\{\bar{x}_n\}$ be a particular solution of Eq.(A.1). Then the transformation

$$z_n = 1/(x_n - \bar{x}_n) \tag{A.6}$$

reduces Eq.(A.1) to the linear first-order equation

$$z_{n+1} + \frac{(c_n\bar{x}_n + d_n)^2}{b_nc_n - a_nd_n}z_n + \frac{c_n(c_n\bar{x}_n + d_n)}{b_nc_n - a_nd_n} = 0, \; n = 0, 1, \dots$$

(c) Let $\{u_n\}$ and $\{v_n\}$ be two different (term-by-term) particular solutions of Eq. (A.1). Then the transformation

$$z_n = \frac{1}{x_n - u_n} - \frac{1}{v_n - u_n} \tag{A.7}$$

reduces Eq.(A.1) to the linear homogeneous first-order equation

$$z_{n+1} + \frac{(c_n\bar{x}_n + d_n)^2}{b_nc_n - a_nd_n}z_n = 0, \; n = 0, 1, \dots$$

(d) Let $\{u_n\}$, $\{v_n\}$ and $\{w_n\}$ be three different (term-by-term) particular solutions of Eq.(A.1). Then the general solution of Eq.(A.1) is given by

$$\frac{1}{x_n - u_n} - \frac{1}{v_n - u_n} = C\left(\frac{1}{v_n - u_n} - \frac{1}{w_n - u_n}\right)$$

where C is an arbitrary constant.

(e) The transformation

$$x_n = \frac{P_nX_n + Q_n}{R_nX_n + S_n}$$

reduces Eq.(A.1) into the Riccati difference equation

$$X_{n+1} = \frac{A_nX_n + B_n}{C_nX_n + D_n}, \; n = 0, 1, \dots$$

where

$$
\begin{aligned}
A_n &= P_n(Q_{n+1}c_n - S_{n+1}a_n) + R_n(Q_{n+1}d_n - S_{n+1}b_n) \\
&= \\
B_n &= Q_n(Q_{n+1}c_n - S_{n+1}a_n) + S_n(Q_{n+1}d_n - S_{n+1}b_n) \\
&= \\
C_n &= P_n(R_{n+1}a_n - P_{n+1}c_n) + R_n(R_{n+1}b_n - P_{n+1}d_n) \\
&= \\
D_n &= Q_n(R_{n+1}a_n - P_{n+1}c_n) + S_n(R_{n+1}b_n - P_{n+1}d_n).
\end{aligned}
$$

f) Let $\{y_n'\}$ and $\{y_n''\}$ be two linearly independent particular solutions of Eq.(A.4). Then the general solution of Eq.(A.1) is given by

$$x_n = \frac{y_{n+1}' + Cy_{n+1}''}{y_n' + Cy_n''} \qquad (A.8)$$

where C is an arbitrary constant.

g) The transformation

$$w_n = (c_n x_n + d_n)\alpha_n \qquad (A.9)$$

where $\alpha_0 = 1$ and

$$\alpha_n = \prod_{j=0}^{n-1} \left(\frac{c_{j+1}}{c_j} (c_j b_j - a_j d_j) \right)^{(-1)^{n-j-1}} \qquad \text{for} \quad n = 1, 2, \ldots \qquad (A.10)$$

reduces Eq.(A.1) to the so-called canonical Riccati difference equation

$$w_{n+1} = K_n + \frac{1}{w_n} \qquad (A.11)$$

where

$$K_n = \frac{c_n d_{n+1} + c_{n+1} d_n}{c_n} \prod_{j=0}^{n-1} \left(\frac{c_{j+1}}{c_j} (c_j b_j - a_j d_j) \right)^{(-1)^{n-j-1}} \qquad (A.12)$$

Proof. The proofs of (a), (b), (c), (e) and (g) follow immediately by introducing the corresponding transformations. (d) follows from (b) and the fact that $1/(v_n - u_n)$ and $1/(w_n - u_n)$ are linearly independent solutions of the first-order linear equation (A.6). (f) is a consequence of (a) and the fact that $y_n = C_1 y_n' + C_2 y_n''$, where C_1 and C_2 are arbitrary constants, is the general solution of Eq.(A.4). $\square$

From Theorem A.1(a), it follows that the Riccati difference equation is equivalent to a second-order linear equation. Also, instead of the general Riccati equation (A.1), it suffices to consider the canonical Riccati equation (A.11).

Before we discuss the asymptotic behavior of solutions of Eq.(A.1) we need to present the so called Poincaré - Perron Theorem.

Theorem A.2 (The Poincaré-Perron Theorem) Consider the linear difference equation of order k

$$x_{n+k} + p_n^{k-1} x_{n+k-1} + p_n^{k-2} x_{n+k-2} + \cdots + p_n^0 x_n = 0, \ n = 0, 1, \ldots \qquad (A.13)$$

where $\{p_n^0\}$, ..., $\{p_n^{k-1}\}$ *are given sequences such that*

$$\lim_{n \to \infty} p_n^i = q_i \quad \text{for} \quad i = 0, \ldots, k - 1. \tag{A.14}$$

Assume that the absolute values of all roots $\lambda_1, \ldots \lambda_k$ *of the equation*

$$\lambda^k + q_{k-1}\lambda^{k-1} + \cdots + q_1\lambda + q_0 = 0 \tag{A.15}$$

are distinct.

(a) Let $\{x_n\}$ *be a solution of Eq.(A.13). Then there exists an* $i \in \{1, 2, \ldots, k\}$ *such that*

$$\lim_{n \to \infty} \frac{x_{n+1}}{x_n} = \lambda_i.$$

(b) Assume that $p_n^0 \neq 0$ *for all* n. *Then there exist* k *solutions* $\{x_n^1\}$, ..., $\{x_n^k\}$ *of Eq.(A.13) such that*

$$\lim_{n \to \infty} \frac{x_{n+1}^i}{x_n^i} = \lambda_i \quad \text{for} \quad i = 1, \ldots k. \tag{A.16}$$

For a proof of Poincaré-Perron's Theorem, see Geljfond [1] and Milne-Thompson [1]. For some recent results related to the Poincaré-Perron Theorem, see Hall and Trimble [1], Máté and Nevai [1], and Trench [1].

Part (a) of the following theorem is known as Poincaré's theorem while part (b) is known as Perron's theorem.

Theorem A.3 *Assume that (A.2) holds and that the following two limits exist:*

$$\left. \begin{aligned} \lim_{n \to \infty} \left(d_{n+1} + a_n \frac{c_{n+1}}{c_n} \right) = p, \\ \lim_{n \to \infty} \left(b_n c_n - a_n d_n \right) \frac{c_{n+1}}{c_n} = q. \end{aligned} \right\} \tag{A.17}$$

Furthermore, suppose that

$$|\lambda_1| > |\lambda_2|$$

where λ_1 *and* λ_2 *are the two roots of the quadratic equation*

$$\lambda^2 - p\lambda - q = 0.$$

Then

$$\lim_{n \to \infty} x_n = \lambda_1$$

for every solution $\{x_n\}$ *of Eq.(A.1).*

Proof. Consider the linear second-order difference equation (A.4) where p_n and q_n are given by (A.5). From (A.2) we find $q_n \neq 0$ for all integers n. Furthermore from (A.17) it follows that

$$\lim_{n\to\infty} p_n = p, \quad \lim_{n\to\infty} q_n = q$$

and we see that the conditions of the Poincaré–Perron Theorem are satisfied. Thus there exist two particular solutions $\{y'_n\}$ and $\{y''_n\}$ of Eq.(A.4) such that:

$$\lim_{n\to\infty} \frac{y'_{n+1}}{y'_n} = \lambda_1 \quad \text{and} \quad \lim_{n\to\infty} \frac{y''_{n+1}}{y''_n} = \lambda_2. \tag{A.18}$$

From Theorem A.1(e) it follows that the general solution of Eq.(A.1) is given by

$$x_n = \frac{y'_{n+1} + Cy''_{n+1}}{y'_n + Cy''_n}.$$

Observe that

$$
\begin{aligned}
x_n - \lambda_1 &= \frac{y'_{n+1} + Cy''_{n+1}}{y'_n + Cy''_n} - \lambda_1 = \frac{y'_{n+1} + Cy''_{n+1} - \lambda_1(y'_n + Cy''_n)}{y'_n + Cy''_n} \\
&= \frac{\dfrac{y'_{n+1}}{y'_n} - \lambda_1 + C\left(\dfrac{y''_{n+1}}{y''_n} - \lambda_1\right)\dfrac{y''_n}{y'_n}}{1 + C\dfrac{y''_n}{y'_n}}.
\end{aligned}
\tag{A.19}
$$

From (A.18) it follows that

$$\lim_{n\to\infty}\left|\frac{y'_{n+1}}{y'_n}\right| = |\lambda_1|, \quad \lim_{n\to\infty}\left|\frac{y''_{n+1}}{y''_n}\right| = |\lambda_2|$$

and so for every $0 < \epsilon < (\lambda_1 - \lambda_2)/2$ there exists an integer n_0 such that for all $> n_0$,

$$|\lambda_1| - \epsilon < \left|\frac{y'_{n+1}}{y'_n}\right| < |\lambda_1| + \epsilon \quad |\lambda_2| - \epsilon < \left|\frac{y''_{n+1}}{y''_n}\right| < |\lambda_2| + \epsilon.$$

Then for $n > n_0$ we have

$$|y'_n| > |y'_{n_0}|(|\lambda_1| - \epsilon)^{n-n_0}, \quad |y''_n| < |y''_{n_0}|(|\lambda_2| + \epsilon)^{n-n_0}$$

and

$$\left|\frac{y''_n}{y'_n}\right| < \left|\frac{y''_{n_0}}{y'_{n_0}}\right|\left(\frac{(|\lambda_1| - \epsilon)}{(|\lambda_2| + \epsilon)}\right)^{n-n_0}.$$

Since $|\lambda_1| - \epsilon > |\lambda_2| + \epsilon$ it follows that

$$\lim_{n\to\infty}\left|\frac{y''_n}{y'_n}\right| = \lim_{n\to\infty}\frac{y''_n}{y'_n} = 0$$

and by letting $n \to \infty$ into (A.19) we obtain

$$\lim_{n \to \infty} |x_n - \lambda_1| = 0.$$

The proof of is complete.   □

A thorough analysis of the behavior of solutions of Riccati difference equations with constant coefficients was presented by Brand [1] who established the following result.

Theorem A.4 *Consider the Riccati difference equation with constant coefficients*

$$x_{n+1} = \frac{ax_n + b}{cx_n + d}, \quad n = 0, 1, \dots \tag{A.20}$$

where a, b, c, d *are real numbers such that*

$$ad - bc \neq 0 \quad \text{and} \quad c \neq 0. \tag{A.21}$$

Let λ_1 *and* λ_2 *be the roots of the quadratic equation*

$$\lambda^2 - p\lambda - q = 0$$

where p *and* q *are given by*

$$p = d + a \quad \text{and} \quad q = bc - ad.$$

Then the following statements are true:

(a) If λ_1 *and* λ_2 *are real and*

$$\text{either} \quad |\lambda_1| > |\lambda_2| \quad \text{or} \quad \lambda_1 = \lambda_2$$

then

$$\lim_{n \to \infty} x_n = \lambda_1. \tag{A.22}$$

(b) If $\lambda_1 = \lambda e^{i\phi}$ *and* $\lambda_2 = \lambda e^{-i\phi}$, *with* $\lambda > 0$ *and* $\phi \in (0, \pi]$ *and if* $\phi/\pi = k/m$, *for some integers* $m > 0$ *and* k *with* $(k, m) = 1$, *then every solution of Eq.(A.20) is periodic with minimal period* m.

(c) If $\lambda_1 = \lambda e^{i\phi}$ *and* $\lambda_2 = \lambda e^{-i\phi}$, *with* $\lambda > 0$ *and* $\phi \in (0, \pi]$ *and if* ϕ/π *is irrational, then every solution* $\{x_n\}$ *of Eq.(A.20) is aperiodic and the set of accumulation points of* $\{x_n\}$ *is the set of real numbers.*

Proof. (a) The case where $|\lambda_1| > |\lambda_2|$ is a consequence of Theorem A.3. Hence there only remains to consider the case where $\lambda_1 = \lambda_2$. Then

$$x_n = \frac{(n+1)\lambda_1^{n+1} + C\lambda_1^{n+1}}{n\lambda_1^n + C\lambda_1^n}$$

is a general solution of Eq.(A.20), and so (A.22) is true.

For the proofs of (b) and (c) observe that the general solution of Eq.(A.20) is given by

$$\begin{aligned}
x_n &= \frac{\lambda^{n+1}\sin(n+1)\phi + C\lambda^{n+1}\cos(n+1)\phi}{\lambda^n \sin n\phi + C\lambda^n \cos n\phi} \\
&= \lambda(\cos\phi + \sin\phi\tan(\alpha - n\phi)),
\end{aligned} \tag{A.23}$$

where $C = \cot\alpha$. If $\phi = \frac{k}{m}\pi$, then the sequence

$$\{n\phi \bmod (\pi)\}$$

is periodic and takes m different values. From (A.23) it follows that $\{x_n\}$ is also periodic with minimal period m.

On the other hand, when ϕ is an irrational multiple of π, the sequence

$$\{n\phi \bmod (\pi)\}$$

is aperiodic and dense in the set $(0, \pi]$. Hence the sequence $\{x_n\}$ is also aperiodic with set of accumulation points equal to **R**. □

Next we will consider the following special case of Eq.(A.1),

$$x_{n+1} = \frac{a_n x_n}{1 + c_n x_n}, \quad n = 0, 1, \ldots \tag{A.24}$$

where the sequences $\{a_n\}$ and $\{c_n\}$ are positive and periodic with period p. Eq.(A.24) which is nonautonomous discrete logistic equation with zero delay was investigated by Clark and Gross [1].

Clearly when the initial condition x_0 is positive, every term of the solution $\{x_n\}$ is also positive. The following theorem describes the asymptotic behavior of solutions of Eq.(A.24).

Theorem A.5 *Assume that the sequences $\{a_n\}$ and $\{c_n\}$ are positive and periodic with period p. Then the following statements are true:*

(a) Let

$$\bar{x}_0 = \frac{\displaystyle\prod_{i=0}^{p-1} a_i - 1}{\displaystyle c_0 + \sum_{i=1}^{p-1} (\prod_{j=0}^{i-1} a_j) c_i} \tag{A.25}$$

and let $\{\bar{x}_n\}$ be the solution of Eq.(A.24) with initial condition $\bar{x}_0$. Then $\{\bar{x}_n\}$ is periodic with period p.

(b) If

$$\prod_{i=0}^{p-1} a_i \leq 1 \tag{A.26}$$

then the zero solution of Eq.(A.24) is a global attractor of all positive solutions of the equation.

(c) If

$$\prod_{i=0}^{p-1} a_i > 1 \tag{A.27}$$

then $\{\bar{x}_n\}$ is a globally asymptotically stable periodic solution of Eq.(A.24). More precisely, for every positive solution $\{x_n\}$ of Eq. (A.24) with $x_0 \neq \bar{x}_0$ we have

$$(x_n - \bar{x}_n)(x_0 - \bar{x}_0) > 0 \quad \text{for} \quad n = 0, 1, \ldots \tag{A.28}$$

and

$$|x_n - \bar{x}_n| \leq \frac{K}{\displaystyle\prod_{i=0}^{n-1} a_i} \quad \text{for} \quad n = 0, 1, \ldots \tag{A.29}$$

where K is a positive constant which depends on x_0.

Proof. Since zero is a particular solution of Eq.(A.11), it follows from Theorem A.1(b) that the transformation $x_n = 1/z_n$ reduces Eq.(A.24) to the linear first-order equation

$$z_{n+1} - \frac{1}{a_n} z_n - \frac{c_n}{a_n} = 0, \ n = 0, 1, \ldots . \tag{A.30}$$

The solution of Eq.(A.30) with initial condition z_0 is given by

$$z_n = z_0 \left(\prod_{i=0}^{n-1} a_i^{-1} \right) + \sum_{i=0}^{n-1} c_i \left(\prod_{j=i}^{n-1} a_j^{-1} \right) \quad \text{for} \quad n = 1, 2, \ldots .$$

Hence

$$x_n = \frac{\prod_{i=0}^{n-1} a_i}{\frac{1}{x_0} + c_0 + \sum_{i=0}^{n-1} c_i \left(\prod_{j=0}^{i-1} a_j \right)} \quad \text{for} \quad n = 1, 2, \ldots \tag{A.31}$$

is an explicit solution of Eq.(A.24).

a) Setting $x_0 = \bar{x}_0$ and $n = p$ into (A.31) we obtain $\bar{x}_p = \bar{x}_0$. Since the sequences $a_n\}$ and $\{c_n\}$ are periodic with period p it follows that

$$\bar{x}_{p+1} = \frac{a_{p+1} \bar{x}_p}{1 + c_{p+1} \bar{x}_p} = \frac{a_1 \bar{x}_0}{1 + c_1 \bar{x}_0} = \bar{x}_1.$$

By continuing in a similar fashion we can prove that $\{\bar{x}_n\}$ is periodic with period p.

b) From (A.31) and the periodicity of $\{a_n\}$ and $\{b_n\}$ it follows that for $k = 0, 1, 2, \ldots$ and $j = 0, \ldots k - 1$,

$$x_{kp+j} = \frac{\left(\prod_{i=0}^{p-1} a_i \right)^k \left(\prod_{i=0}^{j-1} a_i \right)}{\frac{1}{x_0} + c_0 + \sum_{i=0}^{kp+j-1} c_i \left(\prod_{m=0}^{i-1} a_m \right)}.$$

First consider the case where

$$\prod_{i=0}^{p-1} a_i < 1.$$

Then

$$x_{kp+j} < x_0 \left(\prod_{i=0}^{p-1} a_i \right)^k \left(\prod_{i=0}^{j-1} a_i \right)$$

and so

$$\lim_{k \to \infty} x_{kp+j} = 0 \quad \text{for} \quad j = 0, \ldots k - 1.$$

Hence $\{x_n\}$ converges to zero.

Next assume that

$$\prod_{i=0}^{p-1} a_i = 1.$$

Then

$$x_{kp+j} < \frac{\prod_{i=0}^{j-1} a_i}{\dfrac{1}{x_0} + (k+1)c_0}$$

and in a way similar to the above we see that $\{x_n\}$ converges to zero.

(c) Since (A.27) is satisfied we have that $\{\bar{x}_n\}$ is a positive periodic solution of Eq.(A.24) with period p. Our goal is to prove that

$$\lim_{n \to \infty} (x_n - \bar{x}_n) = 0 \qquad (A.32)$$

where $\{x_n\}$ is any positive solution of Eq.(A.24). If $x_0 = \bar{x}_0$, then $x_n = \bar{x}_n$ for $n = 0, 1, \ldots$ and (A.32) is proved.

Now assume $x_0 \neq \bar{x}_0$. By using the fact that the zero sequence and $\{\bar{x}_n\}$ are both solutions of Eq.(A.24), it follows from Theorem A.1(c) that the sequence

$$z_n = \frac{1}{x_n} - \frac{1}{\bar{x}_n} \quad \text{for} \quad n = 0, 1, \ldots$$

satisfies the equation

$$z_{n+1} = \frac{z_n}{a_n}, \quad n = 0, 1, \ldots.$$

Then

$$\frac{1}{x_n} - \frac{1}{\bar{x}_n} = \left(\frac{1}{x_0} - \frac{1}{\bar{x}_0} \right) \left(\prod_{i=0}^{n-1} a_i^{-1} \right) \quad \text{for} \quad n = 1, 2, \ldots \qquad (A.33)$$

and by using the periodicity of the sequence $\{a_n\}$ we see that for $k = 0, 1, 2, \ldots$ and $j = 0, \ldots, k-1$

$$\frac{1}{x_{kp+j}} - \frac{1}{\bar{x}_{kp+j}} = \left(\frac{1}{x_0} - \frac{1}{\bar{x}_0} \right) \left(\prod_{i=0}^{p-1} a_i^{-1} \right)^k \prod_{i=0}^{j-1} a_i^{-1}.$$

Since

$$\prod_{i=0}^{p-1} a_i^{-1} < 1,$$

we see that for $j = 0, \ldots, k-1$

$$\lim_{k \to \infty} \left(\frac{1}{x_{kp+j}} - \frac{1}{\bar{x}_{kp+j}} \right) = 0$$

and so

$$\lim_{n \to \infty} \left(\frac{1}{x_n} - \frac{1}{\bar{x}_n} \right) = 0 \qquad (A.34)$$

rom which (A.32) follows. Also, from (A.34) it follows that $\{x_n\}$ is bounded away rom zero and infinity by positive constants.

Set

$$\left.\begin{array}{rcl} m & = & \inf\limits_{n\geq0}\{\bar{x}_n\} = \min\limits_{0\leq i\leq p-1}\bar{x}_i, \\[2mm] M & = & \sup\limits_{n\geq0}\{\bar{x}_n\} = \max\limits_{0\leq i\leq p-1}\bar{x}_i. \end{array}\right\} \tag{A.35}$$

Since $\{\bar{x}_n\}$ is positive and periodic we have

$$0 < m \leq M < \infty.$$

Let $\epsilon_0 \in (0, 1/M)$ be given. Then there exists an integer n_0 such that

$$\left|\frac{1}{x_n} - \frac{1}{\bar{x}_n}\right| < \epsilon_0 \quad \text{for} \quad n \geq n_0$$

and so

$$\frac{1}{M} - \epsilon_0 \leq \frac{1}{\bar{x}_n} - \epsilon_0 < \frac{1}{x_n} \leq \frac{1}{\bar{x}_n} + \epsilon_0 \leq \frac{1}{m} + \epsilon_0 \quad \text{for} \quad n \geq n_0.$$

Hence

$$C \leq x_n \leq D \quad \text{for} \quad n = 0, 1, \dots \tag{A.36}$$

where

$$C = \min\left\{\min_{0\leq i\leq n_0}\{x_i\}, \frac{1}{1/m + \epsilon_0}\right\} > 0,$$

$$D = \max\left\{\max_{0\leq i\leq n_0}\{x_i\}, \frac{1}{1/M - \epsilon_0}\right\} < \infty.$$

From (A.33) and by using (A.35) and (A.36) we obtain

$$\begin{array}{rcl} |x_n - \bar{x}_n| & = & |x_0 - \bar{x}_0| \dfrac{x_n\bar{x}_n}{x_0\bar{x}_0} \displaystyle\prod_{i=0}^{n-1} a_i^{-1} \\[5mm] & \leq & |x_0 - \bar{x}_0| \dfrac{DM}{Cm} \displaystyle\prod_{i=0}^{n-1} a_i^{-1} \end{array}$$

from which the results follow. $\qquad\qquad\qquad\qquad\qquad\qquad\qquad\qquad\square$

Remark A.1 Riccati difference equations appear in Mathematical Biology. For example, the discrete logistic model without delay is a Riccati difference equation. See Clark and Gross [1] and Pielou [1, 2]. For other applications in Mathematical Biology, see Cushing [1]. They have also applications in Optics (see Saaty [1]), Probability Theory (see Conolly and Choo [1]), and Chemistry (see Miroljubov and Soldatov [1]). Some properties of Riccati difference equations are studied in Brand [1], Kečkić [1], Mitrinović and Adamović [1], and Saaty [1].

Appendix B

A Generalized Contraction Principle

In this appendix we will prove a global attractivity result for a higher order system which presents a generalization of the well-known Banach contraction principle.

We first give a brief review of some known results and concepts from matrix algebra that will be useful in this appendix.

a) Let $A = [a_{ij}]_{p \times q}$ and $C = [c_{ij}]_{p \times q}$ be matrices. Then

$$|A| = [|a_{ij}|]_{p \times q}$$

and

$$A \leq C \text{ if and only if } a_{ij} \leq c_{ij} \text{ for } i = 1, \ldots, p \text{ and } j = 1, \ldots, q.$$

b) Let $\{B_n\}$, where $B_n = [b_n^i]_{p \times 1}$, is a sequence of nonnegative vectors in $\mathbf{R}^p$. Then the series $\sum_{n=0}^{\infty} B_n$ converges if and only if for each $i = 1, \ldots, p$ the series $\sum_{n=0}^{\infty} b_n^i$ converges.

c) Let A be a $p \times p$ matrix. Then the matrix series $\sum_{n=0}^{\infty} A^n$ converges if and only if $|\lambda_j| < 1$ for each eigenvalue λ_j of the matrix A. When the matrix series is convergent then,

$$\sum_{n=0}^{\infty} A^n = (I - A)^{-1},$$

where I is the identity matrix.

Consider the nonautonomous system of difference equations

$$X_{n+k} = F(X_n, \ldots, X_{n+k-1}, n), \quad n = 0, 1, \ldots \tag{B.1}$$

where X_n is a vector in $\mathbf{R}^p$, and where the following hypotheses are satisfied:

189

(H$_1$) *F is a matrix function such that* $F : J^{kp} \times \mathbf{N} \rightarrow J^p$, *where* $J = [a, b]$ *is an interval of real numbers.*

(H$_2$) *For every integer n and for every* $U_1, \ldots, U_{k+1} \in J^p$,

$$|F(U_1, \ldots, U_k, n+1) - F(U_1, \ldots, U_k, n)| \leq \sum_{i=1}^{k} A_i |U_i - U_{i+1}| + B_n \qquad (B.2)$$

where $A_1, \ldots, A_k$ *are nonnegative* $p \times p$ *matrices and where for every integer n,* B_n *is a vector in* $\mathbf{R}^k$ *such that*

$$B_n \geq 0 \quad \text{for} \quad n = 0, 1, \ldots \quad \text{and} \quad \sum_{n=0}^{\infty} B_n \text{ converges.} \qquad (B.3)$$

(H$_3$) $|\lambda_j| < 1$ *for each root of the equation*

$$\det(\lambda I - \sum_{i=1}^{k} A_i) = 0.$$

Theorem B.1 *Consider the nonautonomous system of difference equations (B.1), where* $X_0, \ldots, X_{k-1} \in J^p$ *are arbitrarily chosen. Assume that the hypotheses* (H$_1$) *-* (H$_3$) *are satisfied. Then the following statements are true:*

(a) The sequence of functions $\{F(X, \ldots, X, n)\}$ *converges uniformly on* J^p *to a function* $G : J^p \rightarrow J^p$.

(b) The sequence $\{X_n\}$ *is bounded.*

(c) There exists an unique solution $\bar{X} \in J^p$ *of the system*

$$X = G(X)$$

and also this solution is a global attractor of all solution of the system (B.1).

Proof. (a) By putting $U_0 = \cdots = U_{k+1} = X$ into (B.2) we obtain

$$|F(X, \ldots, X, n+1) - F(X, \ldots, X, n)| \leq B_n.$$

Then by using (B.3) it follows that the sequence $\{F(X, \ldots, X, n)\}$ converges uniformly on J^p. Hence

$$G(X) = \lim_{n \to \infty} F(X, \ldots, X, n)$$

exists and G maps J^p into J^p.

b) Since for every $U_1, \ldots, U_k$ and every integer $n \geq 0$,

$$F(U_1, \ldots, U_k, n) \in J^p = [a, b]^k,$$

follows that the sequence $\{X_n\}$ is bounded.

c) Set

$$D_n = |X_{n+1} - X_n|.$$

Then from (B.1) it follows that

$$D_{n+k+i} = |X_{n+k+i+1} - X_{n+k+i}|$$

$$= |F(X_{n+i+1}, \ldots, X_{n+i+k}, n+1) - F(X_{n+i}, \ldots, X_{n+i+k-1}, n)|$$

$$\leq \sum_{j=1}^{k} A_j |X_{n+i+j} - X_{n+i+j-1}| + B_{n+i}$$

$$= A_1 D_{n+i} + \cdots + A_k D_{n+i+k-1} + B_{n+i}, \quad i = 0, 1, \ldots. \tag{B.4}$$

By adding the preceding inequalities for $i = 0, 1, \ldots, m$ we find

$$C_m = \sum_{i=0}^{m} D_{n+k+i} \leq A_1 \sum_{i=0}^{m} D_{n+i} + \cdots + A_k \sum_{i=0}^{m} D_{n+i+k-1} + \sum_{i=0}^{m} B_{n+i}$$

$$\leq (\sum_{j=1}^{k} A_j) \sum_{i=0}^{m-1} D_{n+k+i} + (\sum_{j=1}^{k} A_j) \sum_{i=0}^{k-1} D_{n+i} + \sum_{i=0}^{m} B_{n+i}$$

$$= (\sum_{j=1}^{k} A_j) C_{m-1} + (\sum_{j=1}^{k} A_j) \sum_{i=0}^{k-1} D_{n+i} + \sum_{i=0}^{m} B_{n+i}$$

$$\leq (\sum_{j=1}^{k} A_j)^2 C_{m-2} + (\sum_{j=1}^{k} A_j + (\sum_{j=1}^{k} A_j)^2) \sum_{i=0}^{k-1} D_{n+i}$$

$$+ (I + \sum_{j=1}^{k} A_j) \sum_{i=0}^{m} B_{n+i}$$

$$\leq (\sum_{j=1}^{k} A_j)^m C_0 + (\sum_{j=1}^{k} A_j + \cdots + (\sum_{j=1}^{k} A_j)^m) \sum_{i=0}^{k-1} D_{n+i}$$

$$+ (I + \sum_{j=1}^{k} A_j + \cdots + (\sum_{j=1}^{k} A_j)^{m-1}) \sum_{i=0}^{m} B_{n+i}.$$

As

$$C_0 = D_{n+k} \leq \sum_{i=0}^{k} D_{n+i},$$

it follows from the above that

$$\sum_{i=0}^{m} D_{n+k+i} \le (I - \sum_{j=1}^{k} A_j)^{-1}(\sum_{i=0}^{k} D_{n+i} + \sum_{i=0}^{m} B_{n+i}). \tag{B.5}$$

As $\{X_n\}$ is bounded, $\{D_n\}$ is also bounded and so

$$D = \limsup_{n \to \infty} D_n$$

exist and is finite. Furthermore, from (B.4) we obtain

$$D \le (\sum_{j=1}^{k} A_j)D,$$

that is

$$(I - \sum_{j=1}^{k} A_j)D = C,$$

where C is a nonpositive vector. Since $(I - \sum_{j=1}^{k} A_j)^{-1}$ exists and is nonnegative, we have

$$D = (I - \sum_{j=1}^{k} A_j)^{-1}B \le 0,$$

which implies

$$D = \lim_{n \to \infty} D_n = 0.$$

Since

$$|X_{n+p} - X_{n+k+m+1}| \le \sum_{i=0}^{m} |X_{n+k+i} - X_{n+k+i+1}| = \sum_{i=0}^{m} D_{n+k+i},$$

where m is positive integer, by letting $n \to \infty$, we obtain

$$\lim_{n \to \infty} |X_{n+p} - X_{n+k+m+1}| = 0.$$

Hence the sequence $\{X_n\}$ converges, that is

$$\lim_{n \to \infty} X_n = \bar{X} \in J^p. \tag{B.6}$$

Now we will prove that $\bar{X}$ is the unique fixed point of the function $G(X)$. By using (B.1) we find,

$$|X_{n+k} - G(\bar{X})| = |F(X_n, \ldots, X_{n+k-1}, n) - G(\bar{X})|$$

$$\le |F(X_n,\ldots,X_{n+k-1},n) - F(X_{n+1},\ldots,X_{n+k-1},\bar{X},n-1)|$$

$$+ |F(X_{n+1},\ldots,X_{n+k-1},\bar{X},n-1)$$

$$- F(X_{n+2},\ldots,X_{n+k-1},\bar{X},\bar{X},n-2)|$$

$$+ \cdots + |F(X_{n+k-1},\bar{X},\ldots,\bar{X},n-k+1) - F(\bar{X},\ldots,\bar{X},n-k)|$$

$$+ |F(\bar{X},\ldots,\bar{X},n-k) - G(\bar{X})|$$

$$\le (\sum_{j=1}^{k} A_j)(\sum_{i=0}^{k-2} |X_{n+i} - X_{n+i+1}| + |X_{n+k-1} - \bar{X}|) + \sum_{i=0}^{k-1} B_{n+i-1}$$

$$+ |F(\bar{X},\ldots,\bar{X},n-k) - G(\bar{X})|.$$

Hence

$$\lim_{n\to\infty} X_{n+k} = G(\bar{X})$$

and $\bar{X}$ is a fixed point of the function $G(X)$. It remains to show that $\bar{X}$ is the unique fixed point of $G(X)$. Assume, for the sake of contradiction, that there exists a vector $\tilde{X} \in J^p$ such that $\tilde{X} \ne \bar{X}$ and $\tilde{X} = G(\tilde{X})$. Then

$$|\bar{X} - \tilde{X}| = |G(\bar{X} - G(\tilde{X})| \le |G(\bar{X}) - F(\bar{X},\ldots,\bar{X},n)|$$

$$+ |F(\bar{X},\ldots,\bar{X},n) - F(\bar{X},\ldots,\bar{X},\tilde{X},n-1)|$$

$$+ \cdots + |F(\bar{X},\tilde{X},\ldots,\tilde{X},n-k+1)F(\tilde{X},\ldots,\tilde{X},n-k)|$$

$$+ |F(\tilde{X},\ldots,\tilde{X},n-k) - G(\tilde{X})|$$

$$\le (\sum_{j=1}^{k} A_j)|\bar{X} - \tilde{X}| + \sum_{i=0}^{k-1} B_{n+i-1}$$

$$+ |G(\bar{X}) - F(\bar{X},\ldots,\bar{X},n)| + |F(\tilde{X},\ldots,\tilde{X},n-k) - G(\tilde{X})|.$$

By letting $n \to \infty$, we obtain

$$|\bar{X} - \tilde{X}| \le (\sum_{j=1}^{k} A_j)|\bar{X} - \tilde{X}|$$

and so $\bar{X} = \tilde{X}$. The proof is complete. $\qquad\square$

Remark B.1 Theorem B.1 is proved in Kocic [3] for sequences in complete metric spaces. In a number of papers Jovanović and Udovičić [1], Kečkić and Kocic [1], Kocic [1 - 4], Marjanović [1], Marjanović and Prešić [1], Prešić [1, 2], and Udovičić [1] similar problems were considered for sequences in complete, metric spaces as well as in other more general spaces.

Appendix C

Global Behavior of Systems of Nonlinear Difference Equations

In this appendix we present some known results about the global behavior of solutions of systems of nonlinear difference equations. There is very little known in this direction and so we hope that this appendix may stimulate interest for further research.

C.1   A Discrete Epidemic Model

The system of nonlinear difference equations

$$\left.\begin{array}{rcl}
S_{n+1} & = & e^{-\alpha I_n} S_n \\[2mm]
I_{n+1} & = & \beta I_n + (1 - e^{-\alpha I_n}) S_n \\[2mm]
R_{n+1} & = & (1 - \beta) I_n + R_n
\end{array}\right\} , n = 0, 1, \ldots \qquad (C.1.1)$$

where

$$\alpha \in (0, \infty) \quad \text{and} \quad \beta \in (0, 1) \qquad (C.1.2)$$

is known as the Kermack-McKendrick model of epidemics. See Kermack and McKendrick [1, 2] and Hoppensteadt [1].

Here S_n represents the number of susceptibles at the nth period, I_n the number of infectives at the nth period, and R_n the number of those who by the nth period have passed through the infectious stage of the disease and have been removed with immunity or died.

We will assume that $S_0, I_0 \in (0, \infty)$ and that $R_0 = 0$. Then clearly S_n, I_n, R_n exist and are positive for all $n = 1, 2, \ldots$. Set $N = S_0 + I_0$. Then it follows from

(C.1.1) that

$$S_{n+1} + I_{n+1} + R_{n+1} = S_n + I_n + R_n \quad \text{for} \quad n = 0, 1, \ldots$$

and so the total population remains constant, that is,

$$S_n + I_n + R_n = N \quad \text{for} \quad n = 0, 1, \ldots. \tag{C.1.3}$$

It is also clear that $\{S_n\}$ is strictly decreasing and $\{R_n\}$ is strictly increasing. Set

$$S_\infty = \lim_{n \to \infty} S_n \quad \text{and} \quad R_\infty = \lim_{n \to \infty} R_n.$$

Then $S_\infty \in [0, \infty)$, $R_\infty \in (0, \infty)$, and by taking limits in the last equation of (C.1.1) we see that

$$\lim_{n \to \infty} I_n = 0.$$

Hence

$$S_\infty + R_\infty = N.$$

Note that the first equation in (C.1.1) implies that

$$S_n = \exp(-\alpha \sum_{i=0}^{n-1} I_i) S_0, \quad n = 0, 1, \ldots$$

which in view of the third equation in (C.1.1) and the hypothesis that $R_0 = 0$ yields

$$S_n = \exp(-\frac{\alpha}{1 - \beta} R_n) S_0, \quad n = 0, 1, \ldots.$$

Therefore

$$S_\infty = \exp(-\frac{\alpha}{1 - \beta}(N - S_\infty)) S_0$$

from which it follows that S_∞ is the unique positive solution in the interval $(0, N)$ of the equation

$$x \exp(-\frac{\alpha}{1 - \beta} x) = S_0 \exp(-\frac{\alpha}{1 - \beta} N).$$

The quotient, $F = S_\infty / S_0$, is a measure of the final size of the epidemic. The following so-called **Threshold Theorem** is a consequence of the above discussion.

Theorem C.1.1 *(Threshold Theorem)*

(a) The susceptible population S_n decreases to a positive value S_∞. Hence, some susceptible will always survive the epidemic.

(b) $F = \exp(-\dfrac{\alpha}{1 - \beta} S_0 (1 + \dfrac{I_0}{S_0} - F))$.

C.2   A Plant-Herbivore System

The following model for the interaction of the apple twig borer (an insect pest of the grape vine) and grapes in the Texas High Plains

$$\left.\begin{array}{rcl} x_{n+1} & = & \dfrac{\alpha x_n}{e^{y_n} + \beta x_n} \\[3mm] y_{n+1} & = & \gamma(x_n + 1)y_n \end{array}\right\}, \; n = 0, 1, \ldots \qquad \text{(C.2.1)}$$

where

$$\alpha \in (1, \infty), \; \beta \in (0, \infty), \; \gamma \in (0, 1) \qquad \text{(C.2.2)}$$

and where x_0 and y_0 are arbitrary positive numbers was developed and studied by Allen, Hannigan and Strauss [1]. See also Allen, Hannigan and Strauss [2] and Allen, Strauss, Thorvilson and Lipe [1].

System (C.2.1) has the equilibrium points

$$E_1 = (0,0) \quad \text{and} \quad E_2 = (\frac{\alpha - 1}{\beta}, 0)$$

and when $\alpha + \beta - \beta/\gamma \geq 1$, it also has the equilibrium point

$$E_3 = (\frac{1}{\gamma} - 1, \; \ln(\alpha + \beta - \frac{\beta}{\gamma})).$$

The linearized system about the equilibrium point E_1 is

$$\left.\begin{array}{rcl} X_{n+1} & = & \alpha X_n \\[3mm] Y_{n+1} & = & \gamma Y_n \end{array}\right\}, \; n = 0, 1, \ldots$$

and because $\alpha > 1$ it is unstable. The linearized system about the equilibrium E_2 is

$$\left.\begin{array}{rcl} X_{n+1} & = & \dfrac{1}{\alpha}X_n + \dfrac{1 - \alpha}{\alpha\beta}Y_n \\[3mm] Y_{n+1} & = & \dfrac{\gamma(\alpha + \beta - 1)}{\beta}Y_n \end{array}\right\}, \; n = 0, 1, \ldots \qquad \text{(C.2.3)}$$

with characteristic roots

$$\lambda_1 = \frac{1}{\alpha} \in (0, 1) \quad \text{and} \quad \lambda_2 = \frac{\gamma(\alpha + \beta - 1)}{\beta}.$$

When

$$\frac{\gamma(\alpha + \beta - 1)}{\beta} < 1, \qquad \text{(C.2.4)}$$

the equilibrium point E_2 is locally asymptotically stable. The following result from Allen, Hannigan and Strauss [1] shows that in this case E_2 is indeed globally asymptotically stable.

Theorem C.2.1 *Assume that (C.2.2) and (C.2.4) hold. Then the equilibrium $E_2 = (\frac{\alpha-1}{\beta}, 0)$ of (C.2.1) is globally asymptotically stable.*

Before we present the proof of Theorem C.2.1 we need the following two lemmas. The first lemma is a comparison result whose proof follows easily by induction. The next lemma is a consequence of Theorem A.5.

Lemma C.2.1 *Assume that f is a nondecreasing function and that $\{u_n\}$ satisfies the difference equation,*

$$u_{n+1} = f(u_n), \ n = 0, 1, \ldots.$$

(a) If $\{v_n\}$ satisfies the inequalities

$$v_{n+1} \leq f(v_n), \ n = 0, 1, \ldots$$

and

$$v_0 \leq u_0,$$

then

$$v_n \leq u_n, \ n = 0, 1, \ldots.$$

(b) If $\{w_n\}$ satisfies the inequalities

$$w_{n+1} \geq f(v_n), \ n = 0, 1, \ldots$$

and

$$w_0 \geq u_0,$$

then

$$w_n \geq u_n, \ n = 0, 1, \ldots.$$

Lemma C.2.2 *Assume that $a \in (1, \infty)$ and $b \in (0, \infty)$. Then every positive solution of the equation*

$$z_{n+1} = \frac{az_n}{1 + bz_n}, \ n = 0, 1, \ldots$$

converges (monotonically) to $(a - 1)/b$.

Proof of Theorem C.2.1 Observe that $x_n, y_n \in (0, \infty)$ for $n = 0, 1, \ldots$ and so

$$x_{n+1} < \frac{\alpha x_n}{1 + \beta x_n}, \quad n = 0, 1, \ldots.$$

follows from Lemmas C.2.1 and C.2.2 that

$$\limsup_{n \to \infty} x_n \leq \frac{\alpha - 1}{\beta}. \tag{C.2.5}$$

Hence, for n sufficiently large, say, for $n \geq N_0$, and by using (C.2.4) it follows that,

$$y_{n+1} = \gamma(x_n + 1)y_n < \gamma(\frac{1 - \gamma}{\gamma} + 1)y_n = y_n.$$

Therefore $\{y_n\}$ decreases to zero. Then for $\epsilon \in (0, \alpha - 1)$ and for n sufficiently large, say for $n \geq N_1 > N_0$,

$$x_{n+1} > \frac{\alpha x_n}{(1 + \epsilon) + \beta x_n}, \quad n \geq N_1.$$

It follows from Lemmas C.2.1 and C.2.2 that

$$\liminf_{n \to \infty} x_n \geq \frac{\alpha - 1 - \epsilon}{\beta}. \tag{C.2.6}$$

As ϵ is arbitrary, it follows from (C.2.5) and (C.2.6) that

$$\lim_{n \to \infty} x_n = \frac{\alpha - 1}{\beta}.$$

The proof is complete. $\qquad\qquad\qquad\qquad\qquad\qquad\qquad\qquad\qquad$ $\square$

C.3 Discrete Competitive Systems

Consider the system

$$\left.\begin{array}{rcl} x_{n+1} & = & x_n f(ax_n + by_n) \\[2mm] y_{n+1} & = & y_n g(cx_n + dy_n) \end{array}\right\}, \quad n = 0, 1, \ldots \tag{C.3.1}$$

where

$$\left.\begin{array}{l} a, b, c, d \in (0, \infty) \\[2mm] f, g \in C[[0, \infty), [0, \infty)] \text{ are decreasing,} \quad \text{and} \\[2mm] f(\bar{x}) = 1, \ g(\bar{y}) = 1 \text{ for some } \bar{x}, \bar{y} \in (0, \infty). \end{array}\right\} \tag{C.3.2}$$

In a series of papers Franke and Yakubu [1 - 3] have investigated various extensions and generalizations of this type of discrete system and have obtain some interesting results.

We also assume that the initial conditions x_0 and y_0 are positive numbers. Then, clearly, the solutions $\{x_n\}$ and $\{y_n\}$ of (C.3.1) exist for all n and are positive.

In (C.3.1), x_n and y_n denote the population densities of two species X and Y, respectively, and the functions f and g are density dependent growth functions. The coefficients a, b, c, d are called the **competition coefficients**.

The following theorem establishes the boundedness of all positive solutions of system (C.3.1).

Theorem C.3.1 *Assume that (C.3.2) holds. Then all positive solutions of system (C.3.1) are bounded.*

Proof. Set
$$F(x,y) = xf(ax + by) \quad \text{and} \quad G(x,y) = yg(cx + dy).$$
Consider the function $H \in C[[0, \infty) \times [0, \infty), [0, \infty) \times [0, \infty)]$ defined by
$$H(x,y) = (F(x,y), G(x,y)) \tag{C.3.3}$$
and the sets of points in $[0, \infty) \times [0, \infty)$ defined by
$$\left. \begin{array}{rcl} N_1 & = & \{(x,y)|ax + by \le \bar{x}, x, y \ge 0\} \\[2mm] N_2 & = & \{(x,y)|cx + dy \le \bar{y}, x, y \ge 0\}. \end{array} \right\} \tag{C.3.4}$$

Clearly, the sets N_1 and N_2 are closed and compact. Furthermore, $N = N_1 \cup N_2$ and $H(N)$ are also compact. Let
$$B = \{(x,y) \mid 0 \le x \le \sup F(N), \ 0 \le y \le \sup G(N)\}.$$

Then $\sup F(N)$ and $\sup G(N)$ are finite and the set B is bounded.

The following cases are possible:
(a) $(x_0, y_0) \in B$. If $(x_0, y_0) \in N$, then clearly, $(x_1, y_1) \in B$. On the other hand, if $(x_0, y_0) \notin N$ then
$$0 \le x_1 = F(x_0, y_0) \le x_0 \quad \text{and} \quad 0 \le y_1 = G(x_0, y_0) \le y_0$$
and we have $(x_1, y_1) \in B$. Hence when $(x_0, y_0) \in B$ then for all $n \ge 0$, $(x_n, y_n) \in B$ and so $\{x_n\}$ and $\{y_n\}$ are bounded.
(b) $(x_0, y_0) \notin B$ and $(x_0, y_0) \in N$. Then clearly $(x_1, y_1) \in B$ and the boundedness of $\{x_n\}$ and $\{y_n\}$ follows from (a).
(c) $(x_0, y_0) \notin B$ and $(x_0, y_0) \notin N$. If for some integer n_0, $(x_{n_0}, y_{n_0}) \in B \cup N$, then the boundedness of $\{x_n\}$ and $\{y_n\}$ follows from (a) or (b). The only remaining case

s when for all n, $(x_n, y_n) \notin B \cup N$. Then, for all n, $x_{n+1} = F(x_n, y_n) \leq x_n$ and $y_{n+1} = G(x_n, y_n) \leq y_n$, and so the sequences $\{x_n\}$ and $\{y_n\}$ are nonincreasing and bounded from above. The proof is complete. $\square$

Next we consider some special cases of (C.3.1) with exponential and rational nonlinearities and give condition for the **extinction of species** X, that is conditions which imply that,

$$\lim_{n \to \infty} x_n = 0.$$

Theorem C.3.2 *Consider the system*

$$\left. \begin{array}{rcl} x_{n+1} &=& x_n \exp(r - (ax_n + by_n)) \\[2mm] y_{n+1} &=& y_n \exp(s - (cx_n + dy_n)) \end{array} \right\} , \quad n = 0, 1, \ldots \qquad (C.3.5)$$

and assume that

$$a, b, c, d, r, s \in (0, \infty), \quad \frac{a}{r} > \frac{c}{s}, \quad \text{and} \quad \frac{b}{r} > \frac{d}{s}. \qquad (C.3.6)$$

Then

$$\lim_{n \to \infty} x_n = 0.$$

Proof. Let $\{x_n\}$ and $\{y_n\}$ be positive solutions of (C.3.5). Set

$$z_n = \frac{x_n^{1/r}}{y_n^{1/s}}, \quad n = 0, 1, \ldots.$$

Then

$$z_{n+1} = z_n \frac{x_{n+1}^{1/r}}{y_{n+1}^{1/s}} = \frac{\exp\left(\dfrac{c}{s}x_n + \dfrac{d}{s}y_n\right)}{\exp\left(\dfrac{a}{r}x_n + \dfrac{b}{r}y_n\right)}, \quad n = 0, 1, \ldots$$

and by using (C.3.6) it follows that

$$z_{n+1} = z_n \frac{\exp\left(\dfrac{c}{s}x_n + \dfrac{d}{s}y_n\right)}{\exp\left(\dfrac{a}{r}x_n + \dfrac{b}{r}y_n\right)} < z_n, \quad n = 0, 1, \ldots. \qquad (C.3.7)$$

Hence $\{z_n\}$ is a decreasing sequence of positive numbers. Let

$$z = \lim_{n \to \infty} z_n.$$

If $z = 0$, then since the sequence $\{y_n\}$ is bounded, it follows that $\lim_{n\to\infty} x_n = 0$ and the theorem is proved. Next, assume that $z > 0$. Since $\{x_n\}$ and $\{y_n\}$ are bounded sequences there exists a sequence of integers $\{n_i\}$, such that

$$\limsup_{n\to\infty} x_n = \lim_{i\to\infty} x_{n_i} = \lambda < \infty \quad \text{and} \quad \lim_{i\to\infty} y_{n_i} = \mu.$$

From (C.3.7) we obtain

$$z_{n_i+1} \le z_{n_i} \exp\left(\left(\frac{c}{s} - \frac{a}{r}\right) x_{n_i} + \left(\frac{d}{s} - \frac{a}{r}\right) y_{n_i}\right), \quad n = 0, 1, \ldots$$

which as i tends to ∞ yields

$$z \le z \exp\left(\left(\frac{c}{s} - \frac{a}{r}\right) \lambda + \left(\frac{d}{s} - \frac{a}{r}\right) \mu\right).$$

Therefore $\lambda = \mu = 0$ and so $\lim_{n\to\infty} x_n = 0$. The proof is complete. $\square$

Theorem C.3.3 *Consider the system*

$$\left.\begin{array}{rcl} x_{n+1} & = & x_n(r + ax_n + by_n)^{-p} \\[2mm] y_{n+1} & = & y_n(s + cx_n + dy_n)^{-q} \end{array}\right\}, \quad n = 0, 1, \ldots \qquad (C.3.8)$$

and assume that

$$a, b, c, d, r, s, p, q \in (0, \infty), \quad r > s, \quad a > c, \quad \text{and} \quad b > d. \qquad (C.3.9)$$

Then

$$\lim_{n\to\infty} x_n = 0.$$

Proof. Set

$$z_n = \frac{x_n^{1/p}}{y_n^{1/q}}, \quad n = 0, 1, \ldots.$$

The rest of the proof is similar to the proof of Theorem C.3.2 and is omitted. $\square$

Finally we present the following result.

Theorem C.3.4 *Consider the system*

$$\left.\begin{array}{rcl} x_{n+1} & = & x_n \exp(r - (ax_n + by_n)) \\[2mm] y_{n+1} & = & y_n(s + cx_n + dy_n)^{-q} \end{array}\right\}, \quad n = 0, 1, \ldots \qquad (C.3.10)$$

and assume that

$$a, b, c, d, r, s, p, q \in (0, \infty) \quad \text{and}$$

$$0 < r < \min\left\{\frac{a}{c}(1 - s), \ \frac{b}{d}(1 - s)\right\}. \right\} \tag{C.3.11}$$

Then

$$\lim_{n \to \infty} x_n = 0.$$

Proof. Set

$$z_n = \frac{x_n}{y_n^\alpha}, \quad n = 0, 1, \ldots$$

where

$$\alpha = \max\left\{\frac{a}{qc}, \frac{b}{qd}\right\}.$$

Then

$$z_{n+1} = z_n h(x_n, y_n), \quad n = 0, 1, \ldots$$

where

$$h(x, y) = (s + cx + dy)^{\alpha q} \exp(r - (ax + by)) \quad \text{for} \quad x, y \geq 0.$$

Note that $0 < h(x, y) < 1$ for $(x, y) \in [0, \infty) \times [0, \infty)$. The rest of the proof is similar to the proof of Theorem C.3.2 and is omitted. $\qquad\Box$

Remark C.3.1 The results in this section are very special cases of more general results by Franke and Yakubu [1 - 3]. For some additional results on discrete systems the reader should consult Hutson, Moran and Vickers [1], Schumacher [1, 2], and the relevant references cited therein.

Bibliography

ADLER, I.

1. Sequences with characteristic number. *Fibonacci Quarterly*, 9:147–162, 1971.

AFTABIZADEH, A. R. AND WIENER, J.

1. Oscillatory properties of first order linear functional differential equations. *Applicable Anal.*, 20:165–187, 1985.

2. Oscillatory and periodic solutions for systems of two first order linear differential equations with piecewise constant arguments. *Applicable Anal.*, 26:327–333, 1988.

AFTABIZADEH, A. R., WIENER, J., AND XU, J. M.

1. Oscillatory and periodic properties of delay differential equations with piecewise constant argument. *Proc. Amer. Math. Soc.*, 99:673–679, 1987.

AGARWAL, R. P.

1. *Difference Equations and Inequalities. Theory, Methods and Applications.* Marcel Dekker Inc., New York, 1992.

ALLEN, L. J. S., HANNIGAN, M. K., AND STRAUSS, M. J.

1. Development and analysis of a mathematical model for a plant–herbivore system. (to appear).

2. Mathematical analysis of a mathematical model for a plant–herbivore system. (to appear).

ALLEN, L. J. S., STRAUSS, M. J., THORVILSON, H. G., AND LIPE, W. N.

1. A preliminary mathematical model of the apple twig borer (Coleoptera: Bostricidae) and grapes on the Texas High Planes. *Ecol. Model.*, 58:369–382, 1991.

ANDREASSIAN, A.

1. Fibonacci sequences modulo m. *Fibonacci Quarterly*, 12:51–64, 1974.

205

ARCIERO, M., LADAS, G., AND SCHULTZ, S. W.

1. Some open problems about the solutions of the delay difference equation $x_{n+1} = \frac{A}{x_n^2} + \frac{1}{x_{n-k}^p}$. (to appear).

BABAI, L. AND LENGYEL, T.

1. A convergence for recurrent sequences with application to the partition lattice. *Analysis*, 12:109–119, 1992.

BEARDON, A. F.

1. *Iteration of Rational Functions*. Graduate Texts in Mathematics 132, Springer-Verlag, Ney York, 1991.

BEDDINGTON, J. R., FREE, C. A., AND LAWTON, J. H.

1. Dynamic complexity in predator prey models framed in difference equations. *Nature*, 255:58–60, 1975.

BERGH, M. O. AND GETZ, W. M.

1. Stability of discrete age-structured and aggregated delay-difference population models. *J. Math. Biol.*, 26:551–581, 1988.

BOOLE, G.

1. *A Treatise on the Calculus of Finite Differences*. G. E. Stechert, New York, 1946.

BORWEIN, D.

1. Solution to problem E 3388[1990,428]. *Amer. Math. Monthly*, 99:69–70, 1992.

BOTSFORD, L. W.

1. Further analysis of Clark's delayed recruitment model. *Bull. Math. Biol.*, 54:275–293, 1992.

BRAND, L.

1. A sequence defined by a difference equation. *Amer. Math. Monthly*, 62:489–492, 1955.

2. *Differential and Difference Equations*. John Wiley and Sons, Inc., New York, 1966.

BROWN, A.

1. A second order non-linear difference equation. (to appear).

2. A third order non-linear difference equation. (to appear).

BURKE, J. R. AND WEBB, W. A.

1. Asymptotic behavior of linear recurrences. *Fibonacci Quarterly*, 19:318–321, 1981.

CAMOUZIS, E., LADAS, G., AND RODRIGUES, I. W.

1. On the rational recursive sequence $x_{n+1} = \frac{a+bx_n^2}{c+x_{n-1}^2}$. (to appear).

CAO, Y. AND LADAS, G.

1. Symmetric periodic solutions of rational recursive sequences. *Rocky Mountain J. Math.* (to appear).

CARVALHO, L. A. V. AND COOKE, K. L.

1. A nonlinear equation with piecewise continuous arguments. *Differential and Integral Equations*, 1:359–367, 1988.

CHANG, D. K.

1. Higher-order sequences modulo m. *Fibonacci Quarterly*, 24:138–139, 1986.

CLARK, C. W.

1. A delayed recruitment model of population dynamics with an application to baleen whale populations. *J. Math. Biol.*, 3:381–391, 1976.

CLARK, M. E. AND GROSS, L. J.

1. Periodic solutions to nonutonomous difference equations. *Math. Biosciences*, 102:105–119, 1990.

COLLET, P. AND ECKMAN, J. P.

1. *Iterated Maps on the Interval as Dynamical Systems*. Birkhäuser, Boston, MA, 1980.

CONOLLY, B. W. AND CHOO, Q. H.

1. The waiting time for generalized correlated queue with exponential demand and service. *SIAM J. Appl. Math.*, 37(2):263–275, 1979.

CONWAY, J. H. AND COXETER, H. S. M.

1. Triangulated polygons and frieze patterns. *Math. Gaz.*, 57(400):87–94, 1973.

2. Triangulated polygons and frieze patterns. *Math. Gaz.*, 57(401):175–183, 1973.

CONWAY, J. H. AND GRAHAM, R. L.

1. On periodic sequences defined by recurrences. (unpublished).

COOKE, K. L., CALEF, D. F., AND LEVEL, E. V.

1. Stability or chaos in discrete epidemic models. In V. Lakshmikantham, editor, *Nonlinear Systems and Applications*, pages 73–93. Academic Press, New York, 1977.

COOKE, K. L. AND GYÖRI, I.

1. Numerical approximation of the solutions of delay differential equations on an infinite interval using piecewise constant arguments. *IMA Preprint Series*, 633, 1990.

COOKE, K. L. AND WIENER, J.

1. Retarded differential equations with piecewise constant delays. *J. Math. Anal. Appl.*, 98:265–294, 1984.

2. An equation alternately of retarded and advanced type. *Proc. Amer. Math. Soc.*, 99:726–732, 1987.

3. Neutral differential equations with piecewise constant argument. *Boll. U. M. I.*, 1-B:321–346, 1987.

COXETER, H. S. M.

1. Frieze patterns. *Acta Arith.*, 18:297–310, 1971.

CRISCI, M. R., JACKIEWICZ, Z., RUSSO, E., AND VECCHIO, A.

1. Stability analysis of discrete recurrence equations of Volterra type with degenerate kernels. *J. Math. Anal. Appl.*, 162:49–62, 1991.

CULL, P.

1. Global stability of population models. *Bull. Math. Biol.*, 43:47–58, 1981.

2. Local and global stability of population models. *Biol. Cybern.*, 54:141–149, 1986.

3. Stability of discrete one-dimensional population models. *Bull. Math. Biol.*, 50:67–75, 1988.

CUSHING, J. M.

1. A strong ergodic theorem for some nonlinear matrix models for the dynamics of structured populations. *Natural Resource Modeling*, 3:331–357, 1989.

DEVANEY, R. L.

1. *An Introduction to Chaotic Dynamical Systems*. Addison-Wesley, Menlo Park, 1985.

DEVAULT, R., KOCIC, V. L., AND LADAS, G.

1. Global stability of a recursive sequence. *Dynamic Systems and Appl.*, 1:13–21, 1992.

RIVER, R. D., LADAS, G., AND VLAHOS, P. N.

1. Asymptotic behavior of delay difference equations. *Proc. Amer. Math. Soc.*, 115:105–112, 1992.

ROZDOWICZ, A. AND J. POPENDA, J.

1. Asymptotic behavior of the solutions of the second order difference equation. *Proc. Amer. Math. Soc.*, 99:135–140, 1987.

DELSTEIN–KESHET, L.

1. *Mathematical Models in Biology.* The Random House/Birkhauser Mathematical Series, New York, 1988.

HRHART, E.

1. Associated hyperbolic and Fibonacci identities. *Fibonacci Quarterly*, 21:87–86, 1983.

LAYDI, S. N.

1. Periodicity and stability of linear Volterra difference systems. (to appear).

2. Stability of Volterra difference equations of convolution type. (to appear).

LAYDI, S. N. AND KOCIC, V. L.

1. Global stability of a nonlinear Volterra difference systems. Technical Report 6, Trinity University, San Antonio, Texas, 1992.

LAYDI, S. N. AND ZHANG, S.

1. Stability and periodicity of Volterra difference equations of finite delay. Technical Report 2, Trinity University, San Antonio, Texas, 1992.

2. Periodic solutions of linear Volterra difference equations of convolution type. Technical Report 3, Trinity University, San Antonio, Texas, 1992.

NRLICH, A.

1. On the periods of the Fibonacci sequence modulo m. *Fibonacci Quarterly*, 27:11–13, 1989.

RBE, L. H. AND ZHANG, B. G.

1. Oscillation of discrete analogues of delay equations. In *Proceedings of The International Conference on Theory and Applications of Differential Equations*, pages 300–309, Ohio University, 1988.

FISHER, M. E.

1. Stability of a class of delay–difference equations. *Nonlinear Anal. Theory, Methods and Applications*, 8:645–654, 1984.

FISHER, M. E. AND GOH, B. S.

1. Stability results for delayed–recruitment in population dynamics. *J. Math. Biol.*, 19:147–156, 1984.

FISHER, M. E., GOH, B. S., AND VINCENT, T. L.

1. Some stability conditions for discrete-time single species models. *Bull. Math. Biol.*, 41:861–875, 1979.

FRANKE, J. E. AND YAKUBU, A. A.

1. Global attractors in competitive systems. *Nonlinear Anal., Theory, Methods and Applications*, 16:111–129, 1991.

2. Mutual exclusion versus coexistence for discrete competitive systems. *J. Math. Biol.*, 30:161–168, 1991.

3. Geometry of exclusion principles in discrete systems. *J. Math. Anal. Appl.*, 168:385–400, 1992.

FREEDMAN, H. I.

1. *Deterministic Mathematical Models in Population Ecology.* Marcel Dekker Inc., New York, 1980.

GELJFOND, A. O.

1. *Calculus of finite differences.* Nauka, Moskow, 1967.

GEORGIOU, D. A., LADAS, G., AND VLAHOS, P. N.

1. Oscillation and asymptotic behavior of a system of linear difference equations. *Applicable Anal.*, 46:241–248, 1992.

GILL, J. AND MILLER, G.

1. Newton's method and ratios of Fibonacci numbers. *Fibonacci Quarterly*, 19:1–4, 1981.

GOH, B. S. AND AGNEW, T. T.

1. Stability in a harvested population with delayed recruitment. *Math. Biosciences*, 42:187–197, 1978.

GOLUB, G. H. AND ORTEGA, J. M.

1. *Scientific Computing and Differential Equations, An Introduction to Numerical Methods.* Academic Press, Inc., Boston, 1992.

GOPALSAMY, K., KULENOVIĆ, M. R. S., AND LADAS, G.

1. On a logistic equation with piecewise constant argument. *Differential and Integral Equations*, 4:215–222, 1991.

GOPALSAMY, K. AND LADAS, G.

1. On the oscillations and asymptotic behavior of $\dot{N}(t) = N(t)[a + bN(t - \tau) - cN^2(t - \tau)]$. *Quart. Appl. Math.*, 48:433–440, 1990.

GRAHAM, R. L.

1. Problem #1343.*Math. Mag.*,63:125, 1990.

GROVE, E. A., KOCIC, V. L., AND LADAS, G.

1. The Fibonacci sequence modulo π. (to appear).

GROVE, E. A., KOCIC, V. L., LADAS, G., AND LEVINS, R.

1. Oscillation and stability in a simple genotype selection model. *Quart. Appl. Math.*, 1993.

2. Periodicity in a simple genotype selection model. (to appear).

GULICK, D.

1. *Encounters with Chaos.* McGraw–Hill, Inc., New York, 1992.

GURNEY, W. S., BLYTHE, S. P., AND NISBET, R. M.

1. Nicholson's blowflies revisited. *Nature*, 287:17–21, 1980.

GYÖRI, I. AND LADAS, G.

1. Linearized oscillations for equations with piecewise constant arguments. *Differential and Integral Equations*, 2:123–131, 1989.

2. *Oscillation Theory of Delay Differential Equations with Applications.* Clarendon Press, Oxford, 1991.

GYÖRI, I., LADAS, G., AND VLAHOS, P. N.

1. Global attractivity in a delay difference equation. *Nonlinear Anal., Theory, Methods and Applications*, 17:473–479, 1991.

HALE, J. AND KOCAK, H.

1. *Dynamics and Bifurcations, Texts in Applied Mathematics 3.* Springer Verlag, New York, 1991.

HALL, L. M. AND TRIMBLE, S. Y.

1. Asymptotic behavior of solutions of Poincaré difference equations. In *Proceedings of The International Conference on Theory and Applications of Differential Equations*, pages 412–416, Ohio University, 1988.

HAUTUS, M. L. J. AND BOLIS, T. S.

1. Solution to problem E 2721[1978,496]. *Amer. Math. Monthly*, 86:865–866, 1979.

HOOKER, J. W. AND PATULA, W. T.

1. A second order nonlinear difference equation: oscillation and asymptotic growth *J. Math. Anal. Appl.*, 91:9–29, 1983.

HOPPENSTEADT, F. C.

1. *Mathematical Models of Population Biology.* Cambridge University Press, Cambridge, 1982.

HUANG, Y. K.

1. A nonlinear equation with piecewise constant argument. *Applicable Anal.*, 33:183–190, 1989,

HUANG, Y. N.

1. A counterexample of P. Cull's theorem. *Kexue Tonbao*, 31:1002–1003, 1986.

2. A theorem on global stability for discrete population models. *Math. Theory Pract.*, 1:42–43, 1987.

3. A note on stability of discrete population models. *Math. Biosciences*, 95:189–198, 1989.

4. A note on global stability for discrete one-dimensional population models. *Math Biosciences*, 102:121–124, 1990.

HUTSON, V., MORAN, W., AND VICKERS, G. T.

1. On the criterion for survival of species in models governed by difference equations. *J. Math. Biol.*, 18:89–90, 1983.

HUTSON, V. AND SCHMITT, K.

1. Persistence and the dynamics of biological systems. *Math. Biosciences*, 111:1–71, 1992.

IVANOV, A.

1. On global stability in nonlinear discrete model. *Nonlinear Anal., Theory, Methods and Applications.*(to appear).

JAROMA, J. H.

1. Rational recursive sequences. (to appear).

JAROMA, J. H., KOCIC, V. L., AND LADAS, G.

1. Global asymptotic stability of a second-order difference equation. In J. Wiener and J. Hale, editors, *Partial Differential Equations*, pages 80–84. Pitman Research Notes in Mathematics Series, Longman Scientific and Technical, Essex, 1992.

JAROMA, J. H., KURUKLIS, S. A., AND LADAS, G.

1. Oscillations and stability of a discrete delay logistic model. *Ukrain. Math. J.*, 43:734–744, 1991.

JIANSHE, Y. AND XULI, H.

1. Private communication.

JORDAN, C.

1. *Calculus of Finite Differences*. Chalsea, New York, 1950.

JOVANOVIĆ, B. AND UDOVIČIĆ, E.

1. Remark on the convergence of the sequences defined by certain difference equations. *Mat. Vesnik*, 9:227–236, 1972.

JURY, E. I.

1. The inners approach to some problems of system theory. *IEEE Trans. Automat. Contr.*, AC-16:233-240, 1971.

KARAKOSTAS, G., PHILOS, CH. G., AND SFICAS, Y. G.

1. The dynamics of some discrete populations models. *Nonlinear Anal., Theory, Methods and Applications*, 17:1069–1084, 1991.

KEČKIĆ, J. D.

1. Riccati's difference equation and a solution of the linear homogenous second order difference equation. *Math. Balkanica*, 8:145–146, 1978.

KEČKIĆ, J. D. AND KOCIC, V. L.

1. On the convergence of certain sequences, V. *Publ. Inst. Math. (Beograd)*, 18:103–106, 1975.

KELLEY, W. G. AND PETERSON, A. C.

1. *Difference Equations, An Introduction with Applications.* Academic Press, New York, 1991.

KERMACK, W. O. AND MCKENDRICK, A. G.

1. Contributions to the mathematical theory of epidemics, part I. *Proc. Roy. Soc. Ser. A*, 115:700–721, 1927.

2. Contributions to the mathematical theory of epidemics, part II. *Proc. Roy. Soc. Ser. A*, 138:55–83, 1932.

KIVENTIDIS, T.

1. Positive solutions of integrodifferential and difference equations with unbounded delay. (to appear).

KNUTH, D. E.

1. *The Art of Computer Programming. Vol I: Fundamental Algorithms.* Addison-Westly, Reading, Mass., 1968.

KOCIC, V. L.

1. A note on the convergence of certain sequences. *Univ. Beograd. Publ. Elektrotehn. Fak. Ser. Mat. Fiz.*, 486:173–177, 1974.

2. On the convergence of sequences defined by recurrent relations. *Math. Balkanica*, 5:155–162, 1975.

3. A theorem on the convergence of sequences defined by recurrent relations. *Univ Beograd. Publ. Elektrotehn. Fak. Ser. Mat. Fiz.*, 525:149–152, 1975.

4. A few remarks on a previous paper regarding the convergence of certain sequences. *Univ. Beograd. Publ. Elektrotehn. Fak. Ser. Mat. Fiz.*, 583:38. 1977.

KOCIC, V. L. AND LADAS, G.

1. Oscillation and global atrractivity in the discrete model of Nicholson's blowflies *Applicable Anal.*, 38:21–31, 1990.

2. Linearized oscillations for difference equation. *Hiroshima Math. J.*, 22:95–102, 1992.

3. Global attractivity in nonlinear delay difference equations. *Proc. Amer. Math. Soc*, 115:1083–1088, 1992.

4. Global attractivity in second-order nonlinear difference equations. *J. Math. Anal. Appl*, 1993. (to appear).

5. Permanence and global attractivity in nonlinear difference equations. (to appear).

6. On rational recursive sequences. (to appear).

KOCIC, V. L., LADAS, G., AND RODRIGUES, I. W.

1. On rational recursive sequences. *J. Math. Anal. Appl*, 1993. (to appear).

KULENOVIĆ, M. R. S., LADAS, G., AND SFICAS, Y. G.

1. Global attractivity in Nicholson's blowflies. *Applicable Anal.* (to appear).

KURSHAN, R. P. AND GOPINATH, B.

1. Recursively generated periodic sequences. *Can. J. Math.*, 26:1356–1371, 1974.

KURUKLIS, S. A.

1. On a theorem of Levin and May concerning the stability of a delay difference equation. Ph.D. Thesis, University of Rhode Island, 1990.

KURUKLIS, S. A. AND LADAS, G.

1. Oscillation and global attractivity in a discrete delay logistic model. *Quart. Appl. Math.*, 50:227–233, 1992.

LADAS, G., PHILOS, CH. G., AND SFICAS, Y. G.

1. Existence of positive solutions for certain difference equations. *Utilitas Mathematica* (to appear).

LAKSHMIKANTHAM, V., MATROSOV, V. M., AND SIVASUNDARAM, S.

1. *Vector Liapunov functions and stability analysis of nonlinear systems.* Kluwer Academic Publishers, Dordrecht, 1991.

LAKSHMIKANTHAM, V. AND TRIGIANTE, D.

1. *Theory of Difference Equations: Numerical Methods and Applications.* Academic Press Inc., New York, 1988.

LASALLE, J. P.

1. *The Stability of Dynamical Systems*. Society for Industrial and Applied Mathematics, Pensylvania, 1976.

LAUWERIER, H.

1. *Mathematical Models of Epidemics*. Math. Centrum, Amsterdam, 1981.

LEECH, J.

1. The rational cuboid revisited. *Amer. Math. Monthly*, 84:518–533, 1977.

LEVIN, S. A. AND MAY, R.

1. A note on difference-delay equations. *Theor. Pop. Biol.*, 9:178–187, 1976.

LEVINE, S. H., SCUDO, F. M., AND PLUNKETT, D. J.

1. Persistance and convergence of ecosystems: An analysis of some second order difference equations. *J. Math. Biol.*, 4:171-182, 1977

LEWIS, E. L.

1. *Network Models in Population Biology*, Biomath, Vol. 7. Springer-Verlag, New York, 1977.

LYNESS, R. C.

1. Note 1581. *Math. Gaz.*, 26:62, 1942.

2. Note 1847. *Math. Gaz.*, 29:231, 1945.

MACKEY, M. C. AND GLASS, L.

1. Oscillation and chaos in physiological control systems. *Science*, 197:287–289, 1977.

MAMANGAKIS, S. E.

1. Remarks on the Fibonacci series modulo m. *Amer. Math. Monthly*, 68:648–649, 1961.

MARJANOVIĆ, M.

1. A further extension of Banach's contraction principle. *Proc. Amer. Math Soc.*, 19(2):411–414, 1968.

MARJANOVIĆ, M. AND PREŠIĆ, S. B.

1. Remark on the convergence of sequences. *Univ. Beograd. Publ. Elektrotehn Fak. Ser. Mat. Fiz.*, 155:63–64, 1965.

MAROTTO F. R.

1. Some dynamics of second order unimodal difference schemes. *Nonlinear Anal., Theory, Methods and Applications*, 18:277–286, 1992.

2. Period doubling and quadrupling for some second order difference schemes. *Nonlinear Anal., Theory, Methods and Applications*, 19:229–236, 1992.

MÁTÉ, A. AND NEVAI, P.

1. A generalization of Poincaré's theorem for recurrence equations. (to appear).

MAY, R. M.

1. Biological populations obeying difference equations: stable points, stable cycles, and chaos. *J. Theor. Biol.*, 51:511–524, 1975.

2. Deterministic models with chaotic dynamics. *Nature*, 256:165–166, 1975.

3. Simple mathematical models with very complicated dynamics. *Nature*, 261:459–467, 1976.

4. Mathematical models in whaling and fisheries management. In G. F. Oster, editor, *Some Mathematical Questions in Biology, 13*. American Mathematical Society, Providence, 1980.

5. Course 8: Nonlinear problems in ecology and resource management. In R. H. G. Helleman G. Iooss and R. Stora, editors, *Chaotic Behaviour of Deterministic Systems*. North-Holland Publ. Co., 1983.

6. Chaos and the dynamics of biological populations. *Proc. R. Soc. London*, 413:27–44, 1987.

MAY, R. M., CONWAY, G. R., HASSEL, M. P., AND SOUTHWOOD, T. R. E.

1. Time delays, density–dependence and single–species oscillations. *J. Anim. Ecol.*, 43:747–770, 1974.

MAY, R. M. AND OSTER, G. F.

1. Bifurcations and dynamic complexity in simple ecological models. *Am. Nat.*, 110:573–599, 1976.

MICKENS, R. E.

1. *Difference Equations, Theory and Applications*. Van Nostrand, Rheinhold, 1990.

MILNE-THOMPSON, L. M.

1. *The Calculus of Finite Differences.* Macmillan, London, 1951.

MIRA, C.

1. *Chaotic Dynamics.* World Scientific, Singapore, 1987.

MIROLJUBOV, A. A. AND SOLDATOV, M. A.

1. *Linear homogenous difference equations.* Nauka, Moskow, 1981.

MITRINOVIĆ, D. S. AND ADAMOVIĆ, D. D.

1. *Sequences and Series.* Naucna knjiga, Beograd, 1990.

MORGAN, K. A.

1. The Fibonacci sequence f_m modulo l_m. *Fibonacci Quarterly*, 21:304–305, 1983.

MORIMOTO, Y.

1. Bifurcation diagram of recurrence equation, $x(t + 1) = ax(t)(1 - x(t) - x(t - 1) - x(t - 2))$. *J. Phys. Soc. Japan*, 53:2460–2463, 1984.

2. Variation and bifurcation diagram in difference equation, $x(t + 1) = ax(t)(1 - x(t) - bx(t - 1))$. *Trans. IEICE*, 72:1–3, 1989.

3. Bifurcation diagram of dc–biased delayed regulation model, $x(t+1) = ax(t)(1 - x(t - 1)) + e$. *Phys. Letters A*, 147:199–203, 1990.

MURRAY, J. D.

1. *Mathematical Biology.* Biomathematics 19, Springer-Verlag, New York, 1989.

NAGASAKA, K.

1. Distribution of recursive sequences defined by $u_{n+1} \equiv u_n + u_n^{-1}$ (mod m). *Fibonacci Quarterly*, 22:76–81, 1984.

NICHOLSON, A. J.

1. An outline of the dynamics of animal populations. *Austral. J. Zool.*, 2:9–25, 1954.

NÖRLUND, N. E.

1. *Differenzenrechnung (Vorlesungen über Differenzenrechnung).* Springer - Verlag, Berlin, 1924.

NORRIS, D. O. AND SOEWONO, E.

1. Period doubling and chaotic behavior of solutions to $y'(t) = \mu y(t)(1 - y(\delta[[(t + \alpha)/\delta]]))$. *Applicable Anal.*, 40:181–188, 1991.

ORTHSFIELD, S.

1. Private communication

APANICOLAOU, V. G.

1. On the asymptotic stability of a class of linear difference equations. (to appear).

HILOS, CH. G., PURNARAS, I. K., AND SFICAS, Y. G.

1. Global attractivity in a nonlinear difference equation. (to appear).

IELOU, E. C.

1. *An Introduction to Mathematical Ecology.* Wiley Interscience, New York, 1969.

2. *Population and Community Ecology.* Gordon and Breach, New York, 1974.

REŠIĆ, S. B.

1. Sur la convergence des suites. *C. R. Acad. Sci. Paris*, 260:3828–3830, 1965.

2. Sur une classe d'inéquations aux différences finies et sur la convergence de certaines suites. *Publ. Ins. Math. (Beograd)*, 5:75–78, 1965.

RESTON, C.

1. *Iterates of Maps on an Interval.* Lecture Notes in Mathematics, 999, Springer-Verlag, 1982.

OBINSON, D. W.

1. The Fibonacci matrix modulo *m*. *Fibonacci Quarterly*, 1:29–35, 1963.

ODRIGUES, I. W.

1. Oscillation and attractivity in a discrete model with quadratic nonlinearity. *Applicable Anal.*, 47:45–55, 1992.

OSENKRANZ, G.

1. On global stability of discrete population models. *Math. Biosciences*, 64:227–231, 1983.

AATY, T. L.

1. *Modern Nonlinear Equations.* Mc Graw Hill Inc., New York, 1967.

ANDEFUR, J. T.

1. *Discrete Dynamical Systems, Theory and Applications.* Clarendon Press, Oxford, 1990.

SCHUMACHER, K.

1. No escape regions and oscillations in second order predator-pray reccurences
   J. Math. Biol., 16:221–231, 1983.

2. Erratum: No escape regions and oscillations in second order predator-pray
   reccurences. *J. Math. Biol.*, 18:91–92, 1983.

SEIFERT, G.

1. On an interval map associated with a delay logistic equation with discontinuous
   delay. in *Delay Differential Equations and Dynamical Systems*, Claremont
   1990, 243–249.

2. Certain systems with piecewise constant feedback controls with a time delay
   Differential and Integral Equations, 1993, (to appear).

SILJAK, D. D.

1. Nonnegative polynomials: a criterion. *Proc. IEEE*, 58:1370–1371, 1970.

2. Stability criteria for two-variable polynomials. *IEEE Trans. Circuits Syst.*
   CAS-22:185–189, 1975.

SINGER, D.

1. Stable orbits and bifurcation of maps on the interval. *SIAM J. Appl. Math*
   35:260–267, 1978.

SMITH, J. M.

1. *Models in Ecology.* Cambridge: Cambridge University Press, 1974.

SZAFRANSKI, Z. AND SZMANDA, B.

1. A note on the oscillation of some difference equations. *Fasc. Math.*, 21:57–63
   1990.

2. Oscillatory behaviour of difference equations of second order. *J. Math. Anal*
   Appl., 150:414–424, 1990.

3. Oscillatory properties of solutions of some difference systems. *Radovi Matemat*
   ički, 6:205–214, 1990.

SZMANDA, B.

1. Characterization of oscillation of second order nonlinear difference equations
   Bull. Pol. Acad. Sci. Math., 34:133–141, 1986.

2. Oscillatory behaviour of certain difference equations. *Fasc. Math.*, 21:65–78, 1990.

CHORO, D.

1. Regula falsi and the Fibonacci numbers. *Amer. Math. Monthly*, 70:868, 1963.

CRENCH, W. F.

1. Asymptotic behavior of solutions of Poincaré difference equations. (to appear).

JDOVIČIĆ, E.

1. On the convergence of sequences defined by difference equations. *Mat. Vesnik*, 8:249–260, 1971.

VINCE, A.

1. The Fibonacci sequence modulo n. *Fibonacci Quarterly*, 16:403–407, 1978.

VINSON, J.

1. The relation of the period modulo to the rank of apparition of m in the Fibonacci sequence. *Fibonacci Quarterly*, 1:37–45, 1963.

WADDILL, M. E.

1. Some properties of a generalized Fibonacci sequence modulo m. *Fibonacci Quarterly*, 16:344–353, 1978.

WALL, D. D.

1. Fibonacci series modulo m. *Amer. Math. Monthly*, 67:525–532, 1960.

WATKINSON, A. R.

1. Density–dependence in single–species populations of plants. *J. Theor. Biol.*, 83:345–357, 1980.

WATKINSON, A. R., LONSDALE, W. M., AND ANDREW, M. H.

1. Modelling the population dynamics of an annual plant *sorghum intrans* in the wet–dry tropics. *Journal of Ecology*, 77:162–181, 1989.

WAZEWSKA-CZYZEWSKA, M. AND LASOTA, A.

1. Mathematical problems of the dynamics of the red–blood cells system. *Annals of the Polish Mathematical Society, Series III, Applied Mathematics*, 17:23–40, 1988.

YALAVIGI, C. C.

1. Periodicity of second– and third–order recurring sequences. *Fibonacci Quarterly*, 11:163–165, 1973.

Subject Index

Attractivity
    basin of attraction, 9
    global attractor, 9

Characteristic equation, 10
Competitive Systems, 199
Contraction principle
    Banach, 24
    contraction mapping, 24
    generalized, 189

Delay logistic equation, 75
Delay logistic equation with piecewise
        constant arguments, 123
Delay logistic equation with several de-
        lays, 123
Discrete baleen whale model, 119
Discrete delay logistic model, 75

Emden-Fowler equation, 125
Epidemic, 99, 169, 195
Equilibrium point, 4, 8
    asymptotically stable, 9
    attracting, 14
    globally asymptotically stable, 9
    globally asymptotically stable rela-
        tive to a set, 9
    hyperbolic, 14
    locally asymptotically stable, 9
    locally stable, 9
    repelling, 14
    saddle, 14
    stable, 9

unstable, 9
extinction, 201

Fibonacci sequence
    modulo π, 174
    modulo m, 175
frieze patterns, 133

Haematopoiesis, 111

Jury test, 25

Liapunov function, 9
Linearized equation, 13
Logistic equation, 75
Logistic equation with piecewise con-
        stant argument, 168

Neural networks, 172
Nicholson's blowflies, 105

Orbit, 88
Oscillation, 4
    Nonoscillatory sequence, 4
    Oscillatory sequence, 4
        about $\bar{x}$, 4
        about zero, 4
    Strictly oscillatory sequence, 4
        about $\bar{x}$, 4

Periodic cycle

eight cycle, 141
p–cycle, 15
five cycle, 138
Periodic point, 88
Periodic sequence, 15
minimal period, 15
period, 15
prime period, 15
Permanence, 35
Plant-Herbivore System, 197
Poincaré-Perron Theorem, 179
Population model, 17

Riccati difference equation, 177
Ruth-Hurwitz criterion, 10

Schur-Cohn criterion, 11
Schwarzian derivative, 15
Semicycle, 5
negative, 5, 28
positive, 5, 28
trivial, 5
Simple genotype selection model, 80
global asymptotic stability, 82
oscillation, 81
periodicity, 87
Sink, 14
Source, 14
Stability, 9
asymptotic, 9
global asymptotic, 9
global asymptotic relative to a set,
9
local, 9
local asymptotic, 9
stable manifold, 88
Symmetric periodic sequence, 145
Symmetric sequence, 144

Threshold Theorem, 196

Trivial solution, 5

Volterra difference equation
linear, 169
nonlinear, 170

z-transform, 100

ω-limit set, 88

Author Index

Adamović, D. D., 188, 218
Adler, I., 175, 205
Aftabizadeh, A. R., 168, 205
Agarwal, R. P., 3, 25, 132, 205
Agnew, T. T., 3, 210
Allen, L. J. S., 197, 198, 205
Andreassian, A., 175, 205
Andrew, M. H., 159, 220
Arciero, M., 176, 206

Babai, L., 170, 206
Beardon, A. F., 3, 206
Beddington, J. R., 169, 206
Bergh, M. O., 3, 206
Blythe, S. P., 105, 211
Bolis, T. S., 53, 212
Boole, G., 3, 206
Borwein, D., 74, 206
Botsford, L. W., 119, 206
Brand, L., 2, 3, 182, 188, 206
Brown, A., 3, 4, 173, 206
Burke, J. R., 175, 207

Calef, D. F., 4, 169, 208
Camouzis, E., 74, 176, 207
Cao, Y., 152, 207
Carvalho, L. A. V., 168, 207
Chang, D. K., 175, 207
Choo, Q. H., 188, 207
Clark, C. W., 2, 3, 12, 48, 119, 132, 207
Clark, M. E., 183, 188, 207
Collet, P., 3, 173, 207
Conolly, B. W., 188, 207

Conway, G. R., 3, 217
Conway, J. H., 2, 133, 207, 208
Cooke, K. L., 4, 13, 168, 169, 207, 208
Coxeter, H. S. M., 2, 133, 207, 208
Crisci, M. R., 170, 208
Cull, P., 17, 18, 208
Cushing, J. M., 188, 208

Devaney, R. L., 3, 16, 173, 208
DeVault, R., 26, 112, 132, 209
Driver, R. D., 3, 209
Drozdowicz, A., 132, 209

Eckman, J. P., 3, 173, 207
Edelstein–Keshet, L., 3, 209
Ehrhart, E., 175, 209
Elaydi, S. N., 170, 209
Enrlich, A., 175, 209
Erbe, L. H., 132, 209

Fisher, M. E., 3, 4, 18, 53, 119, 122, 132, 210
Franke, J. E., 200, 203, 210
Free, C. A., 169, 206
Freedman, H. I., 3, 210

Geljfond, A. O., 3, 180, 210
Georgiou, D. A., 169, 210
Getz, W. M., 3, 206
Gill, J., 175, 210
Glass, L., 111, 216
Goh, B. S.3, 4, 18, 53, 210
Golub, G. H., 3, 211

Gopalsamy, K., 166, 168, 211
Gopinath, B., 16, 215
Graham, R. L., 152, 207, 211
Gross, L. J., 183, 188, 207,
Grove, E. A., 132, 176, 211
Gulick, D., 3, 211
Gurney, W. S., 105, 211
Györi, I., 3, 5, 13, 25, 100, 105, 111,
    119, 123, 171, 208, 211

Hale, J., 3, 173, 212
Hall, L. M., 180, 212
Hannigan, M. K., 197, 198, 205
Hassel, M. P., 3, 217
Hautus, M. L. J., 53, 212
Hooker, J. W., 125, 131, 132, 212
Hoppensteadt, F. C., 3, 195, 212
Huang, Y. K., 168, 212
Huang, Y. N., 17, 212
Hutson, V., 36, 203, 212, 213

Ivanov, A.,53, 213

Jackiewicz, Z., 170, 208
Jaroma, J. H., 53, 132, 176, 213
Jianshe, Y., 136, 213
Jordan, C., 3, 213
Jovanović, B., 194, 213
Jury, E. T., 25, 213

Karakostas, G., 46, 53, 111, 213
Kelley, W. G., 3, 25, 99, 100, 214
Kermack, W. O., 195, 214
Kečkić, J. D., 188, 194, 213, 214
Kiventidis, T., 170, 214
Knuth, D. E., 175, 214
Kocak, H., 3, 173, 212
Kocic, V. L., 7, 26, 45, 53, 74, 105, 112,
    132, 152, 170, 173, 176, 194,
    209, 211, 213, 214

Kulenović, M. R. S., 105, 168, 211, 215
Kurshan, R. P., 16, 215
Kuruklis, S. A., 12, 132, 213, 215

Ladas, G., 2, 5, 7, 25, 26, 45, 53, 74,
    100, 105, 111, 112, 119, 123,
    132, 152, 166, 168, 169, 170,
    171, 173, 176, 206, 207, 209,
    210, 211, 213, 214, 215
Lakshmikantham, V., 3, 11, 25, 170,
    215
LaSalle, J. P., 3, 9, 216
Lasota, A., 119, 220
Lauwerier, H., 99, 216
Lawton, J. H., 169, 206
Leech, J., 2, 133, 216
Lengyel, T., 170, 206
Level, E. V., 4, 169, 208
Levin, S. A., 3, 4, 12, 216
Levine, S. H., 173, 216
Levins, R., 132, 211
Lewis, E. L., 25, 216
Lipe, W. N., 197, 205
Lonsdale, W. M., 159, 220
Lyness, R. C., 2, 133, 139, 216

Mackey, M. C., 111, 216
Mamangakis, S. E., 175, 216
Marjanović, M., 194, 216
Marotto, F. R., 53, 218
Máté, A., 180, 217
Matorosov, V. M., 170, 215
May, R. M., 3, 4, 12, 80, 88, 119, 173,
    216, 217
McKendrick, A. G., 195, 214
Miller, G., 175, 210
Milne-Thompson, L. M., 3, 180, 218
Mira, C., 3, 218
Miroljubov, A. A., 188, 218
Mitrinović, D. S., 188, 218
Moran, W., 203, 212

Morgan, K. A., 175, 218
Morimoto, Y., 4, 173, 218
Murray, J. D., 3, 218

Nagasaka, K., 175, 218
Nevai, P., 180, 217
Nicholson, A. J., 105, 218
Nisbet, R. M., 105, 211
Nörlund, N. E., 3, 218
Norris, D. O., 168, 218
Nortsfield, S., 157, 176, 219

Ortega, J. M., 3, 211
Oster, G. F., 3, 217

Papanicolaou, V. G., 12, 219
Patula, W. T., 125, 131, 132, 212
Peterson, A. C., 3, 25, 99, 100, 214
Philos, Ch. G., 46, 53, 111, 170, 219
Pielou, E. C., 2, 4, 34, 75, 188, 213, 219
Plunkett, D. J., 173, 216
Popenda, J., 132, 209
Preston, C., 3, 219
Prešić, S. B., 194, 216, 219
Purnaras, I. K., 53, 219

Robinson, D. W., 175, 219
Rodrigues, I. W., 74, 152, 166, 176,
    215, 219
Rosenkranz, G., 17, 219
Russo, E., 170, 208

Saaty, T. L., 2, 55, 188, 219
Sandefur, J. T., 3, 15, 219
Schmitt, K., 36, 213
Schultz, S. W., 176, 206
Schumacher, K., 203, 220
Scudo, F. M., 173, 216
Seifert, G., 132, 220

Sficas, Y. G., 46, 53, 105, 111, 170, 213,
    215, 220
Siljak, D. D., 25, 220
Singer, D., 15, 25, 220
Sivasundaram, S., 170, 215
Smith, J. M., 3, 220
Soewono, E., 168, 218
Soldatov, M. A., 188, 218
Southwood, T. R. E., 3, 217
Strauss, M. J., 197, 198, 205
Szafranski, Z., 132, 220
Szmanda, B., 132, 220

Thoro, D., 175, 221
Thorvilson, H. G., 197, 205
Trench, W. F., 180, 221
Trigiante, D., 3, 11, 25, 215
Trimble, S. Y., 180, 212

Udovičić, E., 194, 213, 221

Vecchio, A., 170, 208, 170, 208
Vickers, G. T., 203, 212
Vince, A., 175, 221
Vincent, T. L., 18, 210
Vinson, J., 175, 221
Vlahos, P. N., 3, 169, 171, 209, 210, 211

Waddill, M. E., 175, 221
Wall, D. D., 175, 221
Watkinson, A. R., 159, 221
Wazewska–Czyzewska, M., 119, 221
Webb, W. A., 175, 207
Wiener, J., 168, 205, 208

Xu, J. M., 168, 205
Xuli, H., 136, 213

Yakubu, A. A., 200, 203, 210

Yalavigi, C. C., 175, 221

Zhang, B. G., 132, 209
Zhang, S., 170, 209